CAPACITY BUILDING THROUGH HERITAGE TOURISM

An International Perspective

Advances in Hospitality and Tourism Series

CAPACITY BUILDING THROUGH HERITAGE TOURISM

An International Perspective

Edited by

Surabhi Srivastava, PhD

APPLE ACADEMIC PRESS

First edition published 2021

Apple Academic Press Inc.
1265 Goldenrod Circle, NE,
Palm Bay, FL 32905 USA
4164 Lakeshore Road, Burlington,
ON, L7L 1A4 Canada

CRC Press
6000 Broken Sound Parkway NW,
Suite 300, Boca Raton, FL 33487-2742 USA
2 Park Square, Milton Park,
Abingdon, Oxon, OX14 4RN UK

Library and Archives Canada Cataloguing in Publication

Title: Capacity building through heritage tourism : an international perspective / edited by Surabhi Srivastava, PhD.

Names: Srivastava, Surabhi, 1978- editor.

Series: Advances in hospitality and tourism book series.

Description: Series statement: Advances in hospitality and tourism book series | Includes bibliographical references and index.

Identifiers: Canadiana (print) 20200329529 | Canadiana (ebook) 2020032960X | ISBN 9781771889100 (hardcover) | ISBN 9781003034391 (ebook)

Subjects: LCSH: Heritage tourism—Case studies. | LCGFT: Case studies.

Classification: LCC G156.5.H47 C37 2021 | DDC 338.4/791—dc23

Library of Congress Cataloging-in-Publication Data

Names: Srivastava, Surabhi, 1978- editor.

Title: Capacity building through heritage tourism : an international perspective / Edited by Surabhi Srivastava, PhD.

Description: First edition. | Palm Bay, FL : Apple Academic Press, [2021] | Series: Advances in hospitality and tourism | Includes bibliographical references and index. | Summary: "Capacity Building Through Heritage Tourism: An International Perspective provides a comprehensive account of the valuable tangible and intangible benefits of the development of heritage tourism. Tourism development is widely acknowledged as a crucial tool to foster the development of rural and urban areas. To this end, this book presents nine case studies from international authors that reflect how tourism development is helpful-economically, socially, and otherwise-for community capacity building. The case studies from the countries of Spain, Portugal, Australia, Dubai, Bangladesh, Nigeria, and India demonstrate the uses of various management strategies and methods for rural and urban areas and cover some of the major topics related to community-based tourism, community capacity building, and community participation in developing heritage tourism. Chapters consider the conservation of heritage resources and tourism promotion of destinations that provide opportunities to local communities to strengthen their economies and social standards. Issues and topics addressed include: Water conservation in urban landscape as natural, cultural, and historic tourism resources Spiritual and religious heritage tourism Cultural tourism and the support of public and private funds Economic development and its effect on cultural and natural resources Public-private-partnership to ensure sustainable development Talent management challenges Tribal tourism and the tribal festivals, which are the mirror of their culture and could be major tourist attractions The methodologies and proposed management strategies discussed by book's researchers and professors will be valuable for policymakers, administrators, tourism promoters, researchers, and academicians who are involved with the tourism industry"-- Provided by publisher.

Identifiers: LCCN 2020040583 (print) | LCCN 2020040584 (ebook) | ISBN 9781771889100 (hardcover) | ISBN 9781003034391 (ebook)

Subjects: LCSH: Heritage tourism--Management--Cross-cultural studies. | Heritage tourism--Economic aspects--Cross-cultural studies | Heritage tourism--Social aspects--Cross-cultural studies.

Classification: LCC G156.5.H47 C37 2021 (print) | LCC G156.5.H47 (ebook) | DDC 910.68--dc23

LC record available at https://lccn.loc.gov/2020040583

LC ebook record available at https://lccn.loc.gov/2020040584

ISBN: 978-1-77188-910-0 (hbk)
ISBN: 978-1-00303-439-1 (ebk)

ABOUT THE ADVANCES IN HOSPITALITY AND TOURISM BOOK SERIES

Editor-in-Chief:
Mahmood A. Khan, PhD
Professor, Department of Hospitality and Tourism Management,
Pamplin College of Business,
Virginia Polytechnic Institute and State University,
Falls Church, Virginia, USA
email: mahmood@vt.edu

This series reports on research developments and advances in the rapidly growing area of hospitality and tourism. Each volume in this series presents state-of-the-art information on a specialized topic of current interest. These one-of-a-kind publications are valuable resources for academia as well as for professionals in the industrial sector.

BOOKS IN THE SERIES:

Food Safety: Researching the Hazard in Hazardous Foods
Editors: Barbara Almanza, PhD, RD, and Richard Ghiselli, PhD

Strategic Winery Tourism and Management: Building Competitive Winery Tourism and Winery Management Strategy
Editor: Kyuho Lee, PhD

Sustainability, Social Responsibility and Innovations in the Hospitality Industry
Editor: H. G. Parsa, PhD
Consulting Editor: Vivaja "Vi" Narapareddy, PhD
Associate Editors: SooCheong (Shawn) Jang, PhD, Marival Segarra-Oña, PhD, and Rachel J. C. Chen, PhD, CHE

Managing Sustainability in the Hospitality and Tourism Industry: Paradigms and Directions for the Future
Editor: Vinnie Jauhari, PhD

Management Science in Hospitality and Tourism: Theory, Practice, and Applications
Editors: Muzaffer Uysal, PhD, Zvi Schwartz, PhD, and Ercan Sirakaya-Turk, PhD

Tourism in Central Asia: Issues and Challenges
Editors: Kemal Kantarci, PhD, Muzaffer Uysal, PhD, and Vincent Magnini, PhD

Poverty Alleviation through Tourism Development: A Comprehensive and Integrated Approach
Robertico Croes, PhD, and Manuel Rivera, PhD

Chinese Outbound Tourism 2.0
Editor: Xiang (Robert) Li, PhD

Hospitality Marketing and Consumer Behavior: Creating Memorable Experiences
Editor: Vinnie Jauhari, PhD

Women and Travel: Historical and Contemporary Perspectives
Editors: Catheryn Khoo-Lattimore, PhD, and Erica Wilson, PhD

Wilderness of Wildlife Tourism
Editor: Johra Kayeser Fatima, PhD

Medical Tourism and Wellness: Hospitality Bridging Healthcare (H_2H)©
Editor: Frederick J. DeMicco, PhD, RD

Sustainable Viticulture: The Vines and Wines of Burgundy
Claude Chapuis

The Indian Hospitality Industry: Dynamics and Future Trends
Editors: Sandeep Munjal and Sudhanshu Bhushan

Tourism Development and Destination Branding through Content Marketing Strategies and Social Media
Editor: Anukrati Sharma, PhD

Evolving Paradigms in Tourism and Hospitality in Developing Countries: A Case Study of India
Editor: Bindi Varghese, PhD

The Hospitality and Tourism Industry in China: New Growth, Trends, and Developments
Editor: Jinlin Zhao, PhD

Labor in Tourism and Hospitality Industry: Skills, Ethics, Issues, and Rights
Abdallah M. Elshaer, PhD, and Asmaa M. Marzouk, PhD

Sustainable Tourism Development: Futuristic Approaches
Editor: Anukrati Sharma, PhD

Tourism in Turkey: A Comprehensive Overview and Analysis for Sustainable Alternative Tourism
Editor: Ahmet Salih İKİZ

Capacity Building Through Heritage Tourism: An International Perspective
Editor: Surabhi Srivastava, PhD

The Hospitality and Tourism Industry in ASEAN and East Asian Destinations: New Growth, Trends, and Developments
Editors: Jinlin Zhao, PhD, Lianping Ren, D.HTM, and Xiangping Li, PhD

Event Tourism in Asian Countries: Challenges and Prospects
Editors: Shruti Arora, PhD

ABOUT THE SERIES EDITOR

Mahmood A. Khan, PhD, is a Professor in the Department of Hospitality and Tourism Management, Pamplin College of Business at Virginia Tech's National Capital Region campus. He has served in teaching, research, and administrative positions for the past 35 years, working at major U.S. universities. Dr. Khan is the author of several books and has traveled extensively for teaching and consulting on management issues and franchising. He has been invited by national and international corporations to serve as a speaker, keynote speaker, and seminar presenter on different topics related to franchising and services management. Dr. Khan has received the Steven Fletcher Award for his outstanding contribution to hospitality education and research. He is also a recipient of the John Wiley & Sons Award for lifetime contribution to outstanding research and scholarship; the Donald K. Tressler Award for scholarship; and the Cesar Ritz Award for scholarly contribution. He also received the Outstanding Doctoral Faculty Award from Pamplin College of Business. He has served on the Board of Governors of the Educational Foundation of the International Franchise Association, on the Board of Directors of the Virginia Hospitality and Tourism Association, as a Trustee of the International College of Hospitality Management, and as a Trustee on the Foundation of the Hospitality Sales and Marketing Association's International Association. He is also a member of several professional associations.

ABOUT THE EDITOR

Surabhi Srivastava, PhD, is currently working as senior researcher and associate professor at the Om Kothari Institute of Management, Kota, Rajasthan, India. Her core research area is exploring heritage tourism and the potential of community development. She completed her doctorate and postdoctoral research on heritage tourism development. She also studied tribal and nature tourism development with alternative tourism strategies. To date, she has authored five books on tourism development and has participated at and presented many research papers at national and international seminars and conferences. She has been teaching tourism management subjects for the last 20 years at various institutes of Lucknow, Kanpur, and Kota, India.

CONTENTS

CONTRIBUTORS

José Alvarez-García holds a PhD in Direction and Planning of Tourism by the University of Vigo (Spain). Currently he is an associate professor and researcher in the Accounting and Financial Economy Department of the Faculty of Business, Finance and Tourism at the University of Extremadura, Spain. Previously he was a professor at the University of Vigo. He has an official university master's in Business Innovation and Entrepreneurship (University of Vigo) and an Executive MBA of Business Administration and Management (Business School Caixa Nova) as well as a bachelor of Veterinary Science from the University of León (Castilla y León, Spain), and a diploma in Health from the National School of Health (Spain). Dr. Alvarez-García is the author of numerous research articles in indexed international and national journals and book chapters, and he is also a book editor with Springer, Routledge, and IGI-Global. His main research topics are quality management systems, rural tourism, sustainability, health and wellness, wine tourism, active aging, public pensions system, entrepreneurship, and teaching innovation. E-mail: pepealvarez@unex.es

Ms. Uchenna Ngozi Arene is a curator in the Department of Archaeology and Tourism, University of Nigeria, Nsukka, Enugu State. She is a doctoral student in the same department and has an interest in museum studies, nature-based attractions, among others. E-mail: areneuche@yahoo.com

Francisco Javier Castellano-Álvarez holds a PhD in Sustainable Territorial Development from the University of Extremadura, Spain, where he is currently an assistant professor and researcher in the Department of Economics of the Faculty of Business, Finance and Tourism of Cáceres. He is also author of communications to congresses, book chapters, and academic articles in national and international academic journals. His lines of research are agrarian economy, rural development, and sustainable tourism. E-mail: fcocastellano@unex.es

Ricardo Erasun Cortés is an archaeologist and Managing Partner of Arqueoges–Archeology and Heritage Management, Lda. He is responsible for the scientific direction of more than 100 archaeological interventions in Mainland Portugal and Islands. Since 2000, he was the Scientific Director of archaeological interventions in the Monastery of Santa Maria de Pombeiro de Ribavizela, National Monument, Portugal. He holds a degree in Geography and History, with a specialization in Prehistory and Archeology from the Autonomous University of Madrid, Spain; a master's degree in Cultural Heritage Management from the Universidad Complutense de Madrid; and a dDoctorate in Archeology, History of Antiquity and Historical Sciences and Techniques from the Department of Archeology of the University of Santiago de Compostela, Spain. He also as a degree in Architecture Sciences and is attending his last semester for an integrated master's degree in Architecture and Urbanism at Gallaecia High School, University College. His main scientific interest is the study and interpretation of historical buildings through the detailed graphic records and readings of paramentos, namely the Monastery of Santa Maria de Pombeiro de Ribavizela, which was his object of study for his PhD thesis; as well as the registration, cataloging, and publication of ethnological and archaeological elements linked to the production of bells.

María de la Cruz Del Rio-Rama holds a PhD in Management and Business by the University of Vigo (Spain). Currently she is an associate professor and researcher in the Marketing and Enterprises Organization Department of the University of Vigo. She is author of numerous research articles in international and national journals as well as book chapters. She is also a book editor with Springer, Routledge, and IGI-Global. She was also invited professor in several master programs in the area of quality management. Publications and collaborations in research projects linked with business and tourism are numerous in her career. Her main research topics are quality management systems, rural tourism, sustainability, health and wellness, wine tourism, active aging, entrepreneurship, and teaching innovation. E-mail: delrio@uvigo.es

Prof. Isabel Vaz de Freitas, is an Associate Professor with aggregation at Portucalense University, Porto, Portugal. She has a PhD in History. She is a research member of REMIT (Research on Economics, Management and Information Technologies); research member of LAB2PT at Minho

University; Correspondent of the Portuguese Academy of History; and Head of the Department of Tourism, Heritage and Culture. She is also a coordinator of several study cycles in the area of heritage and tourism and a professor overseeing several undergraduate theses, internships and master's degree reports, master's dissertations, and doctoral dissertation juries, as well as coordinator of national and international conferences in the areas of heritage, culture, and tourism. In addition, she is a collaborator in other projects of valorization of the Territory with City Councils and local organizations. She worked on research projects financed in Portugal and Spain, and in the scope of these projects, she has published articles and books that are centered on themes related to cultural landscape, territorial development and tourism, creative tourism, and water landscape.

Amador Durán-Sánchez is a substitute teacher in the Department of Financial Economics and Accounting, Faculty of Business, Finance and Tourism of Caceres, University of Extremadura, Spain. He is a doctoral student in the program of Sustainable Territorial Development. He obtained his master's degree in Research in Social and Legal Sciences–MUI (specialty: economics and business) and a graduate in Business Administration and Management from the University of Extremadura. He is author of communications to congresses, research articles in national and international journals, and book chapters. His lines of research are water management, rural tourism, active aging, wine tourism, quality and teaching innovation. E-mail: amduransan@unex.es

Dr. Farooq Haq holds a PhD in Marketing from Charles Darwin University, Australia, and an MBA in Marketing from Central Queensland University, Australia. Dr. Haq is among the primary researchers in the area of marketing spiritual tourism, and he is considered as an expert in Islamic marketing. Since 2016, he has been a member of the editorial board for the *Journal of Islamic Marketing*, published byEmerald (UK) and listed by Scopus. In 2018, the *Journal of Islamic Marketing* named him the Outstanding Reviewer. The research interests of Dr. Haq include special interest tourism marketing, Islamic marketing, Halal branding, and medical and heritage tourism marketing.

Md. Shafiqul Islam earned a PhD in Disaster Management from the University of Dhaka, Dhaka, Bangladesh. Dr. Islam is currently working as an Assistant Professor in the Department of Environmental Science,

State University of Bangladesh. He has 23 years of experiences in a cross range of fields. He has conducted a good amount of research in diverse fields including natural resources management (water, forest, land), climate change adaptation, organic farming, sustainable development, renewable energy, and social sciences. His books, articles, and research on immense issues have been published in both national and international journals and books. He has participated and presented his research findings at international and national conferences and workshops. He is interested in the research field of integrated water resource management, disaster management, coastal zone management, waste management, natural resources management, and ecotourism and climate change adaptation.

Dr. Michał Ryszard Koskowski is a researcher, academic teacher, and heritage management consultant specializing in innovative explorations at the intersection of economics, tourism, and heritage studies. He is a graduate of International Trade Studies at the Warsaw School of Economics, Poland, and obtained his PhD in Heritage Management at the International Centre for Cultural and Heritage Studies, Newcastle University, UK. He is employed as a tenure track Assistant Professor at the Collegium of World Economy, Warsaw School of Economics, Poland. He is a member of the Forum UNESCO "University and Heritage" and the Research on Economics, Management and Information Technologies (REMIT) group at Universidade Portucalense.

Shushant Kumar Kushwaha is a PhD Research Scholar in Tourism in the Department of History of Art, Banaras Hindu University, Varanasi, India. He completed graduation in Commerce and holds a diploma in Tourism Management. He earned a master of Tourism Administration from Banaras Hindu University. He also holds the diploma in Polish language and in Integrated Rural Development Program. Apart from education, he served the travel industry as a management trainee in Geanis Holidays and as a business executive at Travel Earth. He was a awarded scholarship in the Polish language to attend a 20-day summer school in Rzeszow University, Poland and was selected to participate in the International Youth Summer School in Chengdu, China.

Paulo Pereira Leite holds a degree in History from the Portucalense University, Portugal. He completed the first year of the master's degree in Modern History, with a specialization in Religious History, at the Faculdade de Letras, Porto, and also completed the first year of the master's degree in Archeology at the University of Minho, Braga, Portugal.

Sudhanshu Mani is a PhD Research Scholar in Tourism Management, Faculty of Arts, Banaras Hindu University, Varanasi, UP, India. After completion of BA (Hons) in French studies, he completed a degree in Tourism Administration from Banaras Hindu University. He worked as management trainee at Via with B2B domestic interface and is now enrolled in a PhD, pursuing a special interest in ICT. He has attended several international and national conferences and training programs.

Anita Medhekar is a senior lecturer in Economics at Central Queensland University, Rockhampton Australia. She has taught economics at the undergraduate and postgraduate levels. Her research interests are in microeconomics and macroeconomics, economic geography and history, international trade and applied economics related to health and tourism economics, development economics, Asia-Pacific economics, poverty and inequality, globalization, privatization, and public-private partnerships. She has presented papers at international conferences and has numerous refereed book chapters and journal articles to her credit.

Chigozie Jude Odum is a lecturer in the Department of Archaeology and Tourism, University of Nigeria Nsukka, Enugu State. He is trained in both humanistic and social sciences. His areas of specialization include cultural resource management, tourism, and ecotourism planning, among others. He has published some articles in both local and international journals. E-mail: jude.odum@unn.edu.ng;odumcj18@yahoo.co.uk

F. O. F. Onwudufor is the current Head of Department of History and International Studies at Nwafor Orizu College of Education, Nsugbe Anambra State, Nigeria. He holds a doctorate degree from the University of Nigeria. He has published articles in both local and international journals. He also has some books published in the area of Igbo proverbs. His areas of specialization include cultural resources management, cultural tourism, Igbo studies, and tourism. E-mail: frfof@ yahoo.co.uk

Makhabbat Ramazanova holds a PhD in Tourism from the University of the Balearic Islands (UIB) in Spain. She holds a master's degree (MSc)

in Sustainable Development from the Dublin Institute of Technology (Ireland) and bachelor's degree in Finance from Sh. Ualikhanov Kokshetau State University (Kazakhstan). Her research interests are sustainable tourism, heritage tourism, water-tourism nexus, water consumption, tourism development impacts, and lake tourism.

Dr. Pravin S. Rana is Assistant Professor of Tourism Management, Faculty of Arts, at Banaras Hindu University Varanasi, India. Dr. Pravin graduated from Banaras Hindu University and holds a master's degree in Tourism Management. He worked on "Pilgrimage and Ecotourism in Varanasi Region: Resources, Perspectives and Prospects" for his doctoral dissertation (University of Lucknow, 2003), and published a dozen papers in this area. He is a co-author of *Banaras Region: A Spiritual and Cultural Guide* (2002), *Tourism Geography* (2006), and *Pilgrimage Tourism in Varanasi Region* (2014). For a few years he had also served as Community Development Officer in International NGO "World Literacy of Canada." He is a life member of the Indian Tourism & Hospitality Congress; INTACH (Indian National Trust for Art, Heritage and Culture), Indian Tourism Congress; and ACLA (Asian Cultural Landscape Association (Korea); and he is an International Fellow at Gifu University, Tokyo, Japan. He has visited Norway and France to deliver lecture on "Religion, Society, Art, Education and Tourism" at Oslo University College, Norway and was invited by World Literacy of Canada at Toronto (Canada). He also visited Malaysia and Singapore to attend the 2nd World Ecotourism Congress. He is coordinating, lecturing, and guiding in various international study programs, such as NYSICCSI: New York State Independent College Consortium for Study in India (USA); Oslo University College Program (Norway); Dragon Program (USA); Heidelberg University study program (Germany); and many others, for more than 15 years. Now he has also joined the team of Planning Varanasi and Varanasi Initiative project from Southern California University, USA. E-mail: psrana@bhu.ac.in

Ummul Fatiha Binte Shafiq is a 4th semester BSc student in Environmental Science and Engineering (ESM) at Jatiya Kabi Kazi Nazrul Islam University, Trishal, Mymensingh, Bangladesh. Her research interest focuses on climate change adaptation, environmental impacts to coastal regions, environmental hazards, hydrological hazards, loss and damage, responsible tourism, and vulnerability analysis and risk reduction. She has participated and presented papers at international conferences.

Dr. Ekta Sharma is an Associate Professor at Amrut Mody School of Management, Ahmedabad University, India, in the area of Organizational Behavior and Human Resource Management. She has teaching experience of more than 16 years. She is on the editorial boards of various reputed national and international journals and has published around 95 research papers in various journals. She is a lead researcher for "High Potential Assessment," U.K. Her areas of interest are individual and interpersonal behavior, heritage management, ethics, and sustainability. She has authored a book titled *Strategic Human Resource Management and Development* with Pearson Higher Education Publishers.

Sandeep Sharma is Director at Magnum Opus International at Ahmedabad, India. He has industry experience of more than 22 years. His first article was published in 2002. His research interests include heritage management, sustainable education, creativity, and innovation. He is working on a project on "Worldwide Creativity Validation for Education and Teachers." His first book on education titled Headmasterji: The Man with Literacy Mission was published in 2016. He is working on sustainable models of the heritage forts of Shekhawati (India) and has published in various journals.

ABBREVIATIONS

BOT	build-operate-transfer
CAP	Common Agricultural Policy
CES	cultural ecosystem services
FGD	Focus Group Discussions
G-PPPs	global-public–private partnerships
IUCN	International Union for Conservation of Nature
ICCROM	International Centre for the Study of the Preservation and Restoration of Cultural Property
LAG	Local Action Group
LEADER	Liaison Entre Actions de Développement de l'Economie Rurale
NTFPs	nonforest timber products
PPPs	public–private partnerships
PRODER	Program of Development and Economic Diversification of Rural Area
REMIT	Research on Economics, Management and Information Technologies
SHT	spiritual heritage tourism
SHTC	spiritual heritage tourism circuits
SIT	special interest tourism
SPA	School of Planning and Architecture
TEK	traditional ecosystem knowledge
UN	United Nations
WHV	World Heritage Volunteers
WHC	World Heritage Council

PREFACE

Capacity building is widely acknowledged as a crucial tool to foster the process of heritage tourism development and considered as a strategy to develop rural/urban areas. This strategy is the essence of "community development" that helps the local community to enhance their skills to participate in tourism activities and increase authentic local knowledge to access external resources.

As the title suggests, this book provides a comprehensive account of tangible and intangible heritage resources for tourism development. It consists of nine well-balanced and logically ordered chapters covering all major topics related to heritage tourism and capacity building. The introduction of the book develops the framework for the analysis and establishes the conceptual background of heritage tourism development for community capacity building. It provides a qualitative literature review of community-based tourism, community capacity building, and community participation in developing heritage tourism in rural and urban areas. It envisages the overview of the research questions based on capacity building through heritage tourism. The book will facilitate this need by providing strategic information and clear insights into the world-wide heritage tourism issues. This conceptual manuscript will help to understand the heritage concept in detail.

Nowadays, water landscape and heritage are at risk due to climate change as an accelerator process, and this continuous degradation of the natural and cultural heritage due to the climate change will affect also the economic opportunities for local communities (UNESCO, 2016). Chapter 2 considers the historic urban landscape, all-natural and cultural attributes placed in historic layers of cultural accumulation, including hydrology and other natural characteristics of the place. In this way, the water presence in historic cities as a natural resource that circulates in rivers, fountains, underground flows, aqueducts, cisterns, and other cultural elements are very important to explore as living testimonies of the human presence and as a natural resource to preserve for the future. The water landscape and the water heritage have a social projection as a legacy and expression of the history of a certain place and the local communities.

This chapter suggested to local governments to involve community members in source water protection planning initiatives. Raising awareness among the visitors on the importance of protection of water would make them feel encouraged to contribute towards this process. This can reflect another part of the capacity building process in the protection of water from both natural and cultural perspectives. In general, it concludes that capacity building plays a significant role in the development, implementation, and maintenance of water resource protection strategies and programs.

Chapter 2 examines the religious spiritual heritage tourism sites and monuments of two neighboring countries, i.e., India and Pakistan, which need to be preserved through capacity building and making people aware of the conservation of this cultural and natural spiritual heritage. The authors had highlighted the need for capacity building between India and Pakistan in partnerships in order to conserve, preserve, and promote, spiritual heritage tourism sites and spiritual cultural heritage which is thousands of years old for economic prosperity, progress and peace between the two countries at all levels people to people, government to government and at the diplomatic level to bring about sustainable transformation between the two countries. Dr. Haq. emphasizes that the inter-generational socio-economic benefits in capacity building for SHT development, conservation, and preservation are far greater for local communities in terms of job creation, growth in national income, foreign exchange earned, than the costs of neglect and destroying centuries old spiritual heritage tourism sites based on religious divide and communal riots.

This chapter brings about an innovative revolution in the travelers to build capacity and drive for heritage conservation and rejuvenation by preparing submissions for world heritage UNESCO listing of the SHT sites in India and Pakistan.

In this diverse series of knowledge, the third chapter throws light on the strategies for rural tourism and the recovery of cultural heritage. The chapter highlights that investments aimed at the promotion of tourism are mainly concentrated on the creation and modernization of accommodation. The authors found the support of public funds for strategies to promote rural tourism in rural areas for private investments with other public actions. This research shows that part of these investments have an exclusive tourism component and aimed at the valorization of various elements of rural heritage that can constitute valuable tourist resources.

The fourth chapter focuses on the conservation of natural heritages with the case study of Sunderban mangroves. Heritage tourism is the safest efforts for the protection of fragile resources as well as makes better socio-economic prospects to the community people. The study concludes that tourist love excitement to enjoy their leisure time for recreation. Sound infrastructure is the basic requirement tourism destination. The natural tourist site may resolve many problems for the community including their livelihoods, natural resources conservation, cultures, values and the spirit of the local community. Cultural ecosystem services and traditional ecosystem knowledge can connect local community and mangroves for the sustainability of culture, heritage and livelihood.

New age tourism is traveling with prior preparations to the destination which is located away from the daily work and new age tourists exhibit the culture, respect the host destination and are environmentally conscious. This concept with the cultural heritage resources is well defined and discussed in the fifth chapter of this book. Dr. Rana explains the dilution of culture and tradition of the heritage site which is being concentrated by organizing. The chapter further discusses the involvement and experience of tourist and residents which makes a suitable example for new age tourism.

As an important economic resource, heritage must be valued and protected. Various risks and challenges are constantly facing heritage, including natural disasters, economic development, tourism, pollution, inappropriate management, plunder, and war. The sixth chapter is emphasizing the issue of economic development which exerts strong pressure on the cultural and natural resources. This quality based study has a focus mainly on the risk factors and heritage stakeholders.

Dr. Odum and other intellects jointly researched and presented the impact of the public private partnership (PPP) on heritage tourism development in the seventh chapter. They applied the mixed method (qualitative and quantitative) to prove that PPP is one of the methods that will be used and accepted by host communities; and these communities believe that it will ensure sustainable development. The chapter illustrates analytically that the PPP approach is economically beneficial, culturally acceptable and environmentally friendly to ensure sustainability.

The eighth chapter of this manuscript is dedicated to the ancient forts and palaces which are converted into heritage hotels and are located in small towns. The study explores the challenges to run and maintain these forts. The efforts to develop, support, restore, and revive the cultural as well as the built heritage of the town and the complexity to get employees for managerial positions, to prepare the community for the involvement for many events become a big task as many small towns are underdeveloped where still the basic facilities are not available to the local community.

The last chapter is the qualitative study of tribal tourism. This study brings many questions and gives several answers for the livelihood and respectable life for the tribes living around the globe. The study prepares a draft and concludes that the tribal festivals are the mirror of their culture and also could be the major attraction for the tourists. It could be the solution of their economic and social problems. Such kind of events is the forced community involvement and core participation. This way they will not feel as commodity or strange.

This book can be evaluated from three different perspectives: a research perspective, an educational perspective, and a tourism policy makers' perspective. In conclusion, given the language of writing and the discussed topics, I envision this book as more suitable for researchers and graduate students, rather than for courses at the undergraduate level. It is a definite must for every university library. The chapters have enormous material for tourism stakeholders, policy makers and are invaluable to the academic world. It comprises many case studies that reflect the capacity building tourism development using different strategies in rural and urban areas.

The chapters of this manuscript also suggested and proposed a new arena for further research for upcoming researchers, conservators, students, in the field of conservation and tourism for international heritage tourism promotion and development.

—Dr. Surabhi Srivastava
Social Scientist, Researcher (Heritage Tourism),
Associate Professor,
Om Kothari Institute of Management Kota,
Rajasthan, India
surabhi3k@gmail.com

ACKNOWLEDGMENT

Heritage tourism development is a process that preserves the living cultures and available natural as well as cultural facts in order to pass on to the future generation. It involves communities people, organizations, civic institutions and governments, all working together. Keeping all the above factors in mind, the present volume is equipped by accumulating all the perspective aspects of heritage tourism.

I am greatly indebted as an editor of this book to the original thinkers and contributors to the subject. I acknowledge with gratitude the contributions of the writers/researchers/thinkers who elucidated the theoretical and practical aspects of heritage tourism that enabled me to compile this manuscript. I deeply bestowed my gratitude to the contributors for sharing their pearls of wisdom and who provided insight and expertise that greatly assisted the research.

I gratefully acknowledge my mentor, Prof. Manoj Dixit, Vice Chancellor, Dr. R.M.L. Awadh University, India. His persistent directions for creativity and knowing things always encouraged me to work for noble and constructive outcomes. His continuous inspiration, supervision and vast knowledge facilitated me to prepare this present volume. I would like to give my special thanks to Dr. M. L. Sahu, Coordinator, Heritage Department, University of Kota, India, for encouraging and motivating my academic efforts. He always continued to be behind my every effort to vitalize me at every step with his valuable suggestions.

I am deeply beholden and a debt of gratitude to Apple Academic Press for making this manuscript in the present shape and especially to Dr. Mahmood A. Khan; without his supervision, guidance, and constant support this book would not have seen light of the day.

Last but not the least, I owe it to my parents, my husband, and my loving daughters, Aanya and Aarna, for their constant support and patience bestowed on me during the long period of my study. Their patience and smiles during my work always encourages me.

I feel very delighted in expressing my gratitude towards my friends, colleagues, and my seniors in creating this book.

—Dr. Surabhi Srivastava

INTRODUCTION

Tourism is a fragmented industry, which means the purchase of one involves the purchase of products from multiple establishments. This is linked to the fact that most of the tourism enterprises are small, medium, and micro enterprises (SMMEs), which means that the tourism industry can benefit the local economy to great depth. The top end of the tourism industry is dominated by large multinationals, with major airlines, hotel groups, and tour operating companies. Local consumption can benefit from tourism consumption, tourism infrastructure can be shared by locals and tourists; moreover, consumption of tourism by tourists can accelerate the delivery of infrastructure to a local, local people can be benefited (Henama, 2017).

"Heritage is what modern society wants to inherit and pass on and history is what a scholar deserves recording" (Tunbridge and Asworth, 1996). According to the Oxford English Dictionary, "Heritage is something that can be passed down from one generation to the next, something that can be preserved or inherited, and has historical or cultural value." Inheritance can be considered a physical "object": a part of a property, a building, or a place that is "capable of ownership," that is able to be "owned," and "passed on" to someone else (Campanella, 2013).

Heritage can become a money-spinner, if it is properly presented to meet the needs of customers who are constantly seeking meaningful experiences (Henama, 2017). Heritage tourism may be an amalgamation of words to appear more widely in tourism planning. This confirms that the local community has identified important elements of their places to be significantly protected and passed on to future generations. Heritage tourism also shows that people are dedicated to caring for and sharing these identified elements or resources.

Preservation of the world's cultural heritage is important for the promotion and education of present and future generations. A large part of tourism depends on places of natural, native, and historical importance which are tourism product based. To honor the cultural significance of the destination, people involved in the tourism industry need to be sensitive to the cultural groups they have a special interest in and need to be directly involved in the planning and promotion of the destination. More importantly, different regions have to recognize that heritage resources are important for their sustainability and are not exogenous factors (Sharma and Sharma, 2017). Heritage tourism can refer to religious tourism, migratory tourism, living culture, historical culture, built heritage, archaeological sites, and ancient monuments. It provides a means of establishing local pride as it emphasizes presenting and accepting local values and cultural memory while allowing local communities to connect with the wider world. Inheritance management can be another challenge.

Tourism authorities-oriented infrastructure development may hesitate to allow the site managers and other authorities to impose any restrictions on the use of the site or to take conservation measures. Interpretation of heritage provides protection of awareness to tourists and locals (Salazar and Xu, 2015).

The understanding and management of heritage tourism is a social phenomenon. It is more subjective in nature and is related to the actual relationship between space and person at the core of the phenomenon. Many researchers presented the challenge that people who come to a place only come to "gaze," are educated or enjoy themselves. For some, it is argued that it is an emotional experience that people "feel" rather than "gaze" (Uruguay 1990).

From this point of view, the industry being an employment-oriented industry has a belief that it can resolve socio-economic anomalies if it is to be extended to many sectors of society. In view of these facts, community-based tourism (CBT) has been started to be developed with broad objectives. It is also argued that there is a need to assess the impact on livelihoods and not just income with a diversity of positive and negative effects. There is a strong belief that CBT engages and empowers native individuals as actors in tourism planning, development, and management. There is a perception among local stakeholders that CBT helps them understand how tourism can contribute to their well-being and can play an active role in reducing the negative effects of tourism. This means that local tourism resources and products should be managed, developed, and owned by local communities themselves to ensure maximum tourism benefits to local stakeholders, making them more responsible.

Capacity development is the process through which capabilities are acquired, strengthened, optimized, and maintained over time. For individuals, competence may relate to leadership, advocacy skills, training/speaking ability, technical skills, organizing skills, and other areas of personal and professional effectiveness. There is a large range of capacity-building approaches—a continuum—that include peer-to-peer learning, organizational development, training and academic studies, research, publications, and grant-building facilities. Along with quality, capacity building (CB) also takes place across organizations, within communities, across geographies, within the nonprofit sector, and in regions. It consists of individuals and groups of individuals, organizations, groups of organizations within the same sector or region, and organizations and actors from different sectors and regions (Whittle et al., 2012).

CB only makes sense when it refers that it has plans to build capacity. It builds the capacity of many individuals who live almost on destination communities, who directly or indirectly lead to initiate and support many social processes. Support a continuous learning society. Social education and empowerment are based on each other. Empowerment is the method of increasing the capacity of individuals or groups and converting selections into desired actions and outcomes.

The relationship between human communities and places is not unique to rural areas, where land becomes part of people's lives, interacting with buying and selling, working and playing, and living and dying. It is both history and future (Swales, 2015).

Community capacity building (CCB) is an essential component for the achievement of community development. Tourism development and CB for indigenous communities are needed for progress at hand. CCB programs facilitate underdeveloped communities to promote their ability to participate in tourism decision-making processes. They encourage community participation and provide tools that enable them to do so.

Capacity development is the method by which individuals, organizations, and societies develop their abilities over time to organize and achieve social and economic goals. It is a plan that extends the period of CB to cover all aspects of creating and maintaining capacity addition over time. It not only includes many forms of teaching and training but also continuous efforts to develop institutions, political awareness, financial resources, technology systems, and a broad enabling environment ("Glossary," United Nations Committee of Experts on Public Administration, 2006).

This confirms that the local community has identified important elements of their places to be significantly protected and passed on to future generations. Heritage tourism also shows that people are dedicated to caring for and sharing these identified elements or resources. To build community capacity, community participants need to acquire skills and knowledge that help people think and act in new ways. The concept of community skills and knowledge is considered to be a tool for tourism development and community development as well as assistance in local communities. Resource mobilization is considered as a means of determining the capacity of the community to identify the dimensions of community capacity as well as the development of tourism resources in local communities. Therefore, mobilizing community resources is an essential component of community capacity. They can affect the ability of the community to achieve their goals. Community resource mobilization comes in many forms (Arif et al., 2010).

Capacity development in communities can be seen as the ability of community residents to participate in tourism activities. An important objective of CB is to verify whether individuals, organizations, and communities have built their capacity for the development of tourism in their communities. Tourism development in local communities cannot be successful without the involvement of community leaders and community residents (Gangnibo, 2010).

CB through tourism development is an essential condition for improving the process of tourism activities and enhancing its benefits for local communities. Tourism development can also be used as a technique to reduce conflicts between conservation and development to increase community participation. Local support and CB is essential prior to tourism development to include native communities. Local communities are usually complex and heterogeneous, and tourism development needs to suit their diverse needs (Chia-Chiubuhui-Mintai, 2014).

Direct or indirect participation of local residents in tourism activities is an important part of developing heritage tourism. There are traces of local influence in the tourism development process (Ernstein, 1969). Local people have control over all development without any external force or influence (Choguil, 1996; Dewar, 1999). Locals can contact explorers directly and develop real tourism by themselves (Choguil, 1996; Dewar, 1999; Sundar, 1995). People are consulted in many ways, for example, attending a community meeting or public hearing. Developers can accept some contributions from local people who take advantage of their project (Ernstein, 1969). The rights of local people are recognized and accepted at the local level (Sundar, 1995).

But, in practice, in most places, people are told about the tourism development program, which has already been decided in the community.

Developers run projects without listening to the opinions of local people (Ernstein, 1969). Tourism development is usually developed by some powerful individuals or government, without any discussion with the people (Ernstein, 1969).

CCB can be broadly defined as "community group growth" to define, analyze, and act on concerns of importance (health, or any other) to its members (Labonet and Leverac, 2001a, p. 16). CCB is additionally seen as a process (Goodman et al., 1998) that increases assets and characteristics that a community is able to attract. CCB is not specific to a particular locality, or to individuals or groups within it, but also to interactions between the two. CB as a technique for sustainable skills, resources, and commitments in various settings has evolved due to the need to increase the profitability of the project (Gibbon, 1927; Labonte and Leverac, 2002). These qualities exist in relation to specific issues and concerns of specific people and groups, and specific activities or projects. Community empowerment and community capacity are closely constructed as social organizations and crowds try to address inequalities in people's lives (Lavark, 2007, p. 19). CCB is often the means by which community empowerment results can be achieved. Both CCB and empowerment are achieved through building knowledge, skills, and competencies at the local level. The vision of CCB is intended to be an empowering experience for communities and involves the use of a participatory "tool" in combination with nine "capability domains" to help people organize themselves and be serious about their individual and collective circumstances. Crucially, it enables people to strategically plan to resolve their situations, to evaluate, and to present the process visually as outcomes. Rather than being a substitute for project objectives, community capacity constitutes a separate set of goals that run "parallel" to the specific purpose of a project. The "tool" is typically implemented as a workshop to build community capacity (Glenn, 2014).

A major challenge in sustainable tourism is to develop economically viable enterprises that provide livelihood benefits to local communities while protecting indigenous cultures and environments. Such undertakings are difficult to assess due to the general lack of effective assessment and monitoring methods, the lack of consensus about the methodology, and the inability of some monitoring systems to include all elements of tourism impacts (Simpson, 2007).

When we talk about CCB, we are basically talking about "local solutions to local problems," which enable communities to deal with problems, ultimately without relying on external resources. CCB can be seen as a way to encourage the development of more "healthy" or "active" communities. The pooling of skills and resources, as well as the identification of key local problems that many seek to address through common action can provide a springboard for effective action in many areas.

A review of the literature about CB and in-depth qualitative studies revealed the relevance of CCB for planning tourism development in rural or urban areas. "Learning opportunities and skills development," "resource mobilization," "partnership relationships/networking," "leadership," "participatory decision making," " asset-based approach," "sense of community," "communication," and "The Development Pathway" are nine broad domains to serve as the foundation for community-based activity by those in the field to support and nurture the development of competent communities (Selma and John, 2011). The findings of this study are that tourism contributes to the literature and community development. In addition, the findings of this study may be useful in assisting tourism planners, and academic researchers to assess, monitor, and evaluate current or potential tourism development. The results of this investigation also assist researchers in the field of community development and the tourism industry. Tourism developers often build community capacity. Community capacities like to encourage a sense of community among residents as a way to contribute to the construction. Thus local communities play an important role in addressing tourism development base CB (Beaton, 2006). In relation to this, community leaders, stockholders, and tourism development practitioners should regard the concept of community spirit as a refinement of community members. While there is a substantial body of literature on the concept of community understanding and definition (Chavis and Wandersman, 1990; Davidson and Couture, 1989; McMillan and Chavis, 1986), only a few discuss the practical application of literary approaches successfully measure a sense of community in various cultural contexts. Local communities need to strengthen the sense of community in order to build community capacity in tourism development (Conway and Hachen, 2005). Lack of a sense of community has led to a full-fledged rumor about explanations for people not participating in development activities (DeNeui, 2003). If the local community is not involved in the conservation process, it is very difficult to protect any heritage site (Sharmin and Khan, 2019). This book illustrates the relevance of CCB and community participation in tourism development in rural and urban areas.

Key points on ensuring good practice in the CCB (extracted from Community Capacity Building–A Practical Guide by Roland Atkinson and Paul Willis, 2006)

- Involve the local people from the start before any action is taken; start with the issues that concern them.
- Locally-based resources are required, such as neighborhood homes, so that support can be readily obtained.
- Changing the existing bureaucratic culture to involve customers in decision-making processes maybe important. Adequate time and resources need to be given for participation and action.
- Local people should be employed where possible in key roles so that management is maintained within the community where possible.
- Making available for a wide variety of small projects rather than a limited number of large projects helps prevent a relatively narrow pool of community members from joining, thereby promoting wider programs and interest within the community.
- It is important to recognize that conflict is a very common part of the CCB process and that community activists facilitating CCB projects require skills to negotiate and resolve small group conflicts.
- There is a need to protect against greater involvement (and potential exploitation) of volunteer community members because the danger of choosing particular community groups and individuals (general suspects) is not excluded from others.
- "Taking it too far"—it may be unfair to push community groups and individuals into CCB projects.
- Community members need to have confidence in the priorities and policies of housing organizations, this can be particularly challenging with housing staff that may be treated with hostility and mistrust by community members and tenants.

REFERENCES

Aref, F. Barriers to Community Capacity Building for Tourism Development in Communities in Shiraz, Iran. *J. Sustain. Tourism* **2011,** *19* (3), 347–359, DOI: 10.1080/09669582.2010.517314.

Aref, F.; Redzuan, M. B. Community Capacity Building for Tourism Development. *J. Hum. Ecol.* **2009,** *27* (1), 21–25, DOI: 10.1080/09709274.2009.11906187.

Atkinson R.; Willis, P. *Community Capacity Building–A Practical Guide*, 2006, http://www.chs.ubc.ca/archives/files/Community%20Capacity-Building%20A%20Practical%20Guide.pdf.

Ghaderi, Z.; Abooali, G.; Henderson, J. Community Capacity Building for Tourism in a Heritage Village: The Case of Hawraman Takht in Iran. *J. Sustain. Tourism* **2018,** *26* (4), 537–550, DOI: 10.1080/09669582.2017.1361429.

Glenn, L. Community Capacity Building. *Encyclopedia of Quality of Life Research*; Michalos, A, Ed.; Springer, 2014, DOI: 10.1007/978-94-007-0753-5_456.

Henama, U. Marikana: Opportunities for Heritage Tourism. *Afr. J. Hospitality Tourism Leisure* **2017,** *6* (4), ISSN: 2223- 814X Copyright: © 2017 AJHTL.

Leksakundilok, A. Community Participation in Ecotourism Development in Thailand. Thesis, University of Sydney, Geosciences, 2006.

Moscardo, G. *Building Community Capacity for Tourism Development.* CABI, 2008, Business & Economics.

Poria, Y.; Butler, R.; Airey, D. The Core of Heritage Tourism. Ann. Tourism Res. **2003,** *30,* 238–254, DOI: 10.1016/S0160-7383(02)00064-6.

Salazar, N.B.; Zhu, Y. Heritage and Tourism. *Global Heritage: A Reader*, 1st ed; Meskell, L. Ed.; John Wiley & Sons, Inc., 2015.

Sharma, A.; Sharma S. Heritage Tourism in India: A Stakeholder's Perspective. *Tourism Travelling (open-access)* **2017,** *1* (1), 20–33, DOI: 10.21511/tt.1(1).2017.03).

Sharmin, S.; Khan, N.M. *Community Based Cultural Tourism: An Attitude towards Sustainable Conservation Model for Heritage Site*, Proceedings of International Conference on Planning, Architecture and Civil Engineering, Feb 07–09 2019; Rajshahi University of Engineering & Technology: Rajshahi, Bangladesh, 2019.

Simpson, M.C. An Integrated Approach to Assess the Impacts of Tourism on Community Development and Sustainable Livelihoods. *Community Dev. J.* **2009,** *44* (2), 186–208, https://doi.org/10.1093/cdj/bsm048.

Svels, K. World Heritage, Tourism and Community Involvement: A Comparative Study of the High Coast (Sweden) and Kvarken Archipelago (Finland). *Scand. J. Hospitality Tourism* **2015,** *15* (1–2), 183–201, DOI: 10.1080/15022250.2015.1009708.

Tunbridge, J. E.; Ashworth, G. J. *Dissonant Heritage: The Management of the Past as a Resource in Conflict*; John Wiley & Sons, 1996.

United Nations Committee of Experts on Public Administration. *Definition of Basic Concepts and Terminologies in Governance and Public Administration (PDF)*; United Nations Economic and Social Council, 2006.

Whittle, S.; Colgan, A.; Rafferty, M. *Capacity Building: What the Literature Tells Us*; The Centre for Effective Services: Dublin, 2012.

Wu, C.-C.; Tsai, H.-M. Capacity Building for Tourism Development in a Nested Social–Ecological System—A Case Study of the South Penghu Archipelago Marine National Park, Taiwan. *Ocean Coastal Manage.* **2016,** *123*, 66–73, ISSN 0964-5691, https://doi.org/10.1016/j.ocecoaman.2016.02.001.

CHAPTER 1

WATER CONSERVATION IN THE URBAN LANDSCAPE THROUGH HERITAGE AND CAPACITY BUILDING: THE CASE OF GUIMARÃES, PORTUGAL

ISABEL VAZ DE FREITAS[1*], MAKHABBAT RAMAZANOVA[2], RICARDO ERASUN CORTÉS[3], and PAULO PEREIRA LEITE[3]

[1]*Universidade Portucalense, Research on Economics, Management and Information Technologies—REMIT, Porto, Portugal*

[2]*University of the Balearic Islands, Palma de Mallorca, Spain*

[3]*Arqueoges—Archeology and Heritage Management, Guimarães, Portugal*

[*]*Corresponding author. E-mail: ifc@upt.pt*

ABSTRACT

Water is vital for our life and for the well-being and survival of future generations. Water is the key element of the socio-economic development of the countries and is one of the most critical resources for the tourism industry. Nowadays water scarcity and water quality challenges have been identified in many parts of the world. Thus, it has been recognized that water is a natural resource, which needs to be preserved and guarantee its quality. In this context, water issues in the tourism industry are obtaining big attention from organizations, such as the United Nations Environmental Programme, the United Nations World Tourism Organization.

Moreover, this resource is an unquestionable element of the urban planning in history, and in the past, the decision-makers assumed its presence as fundamental in the city or rural planning.

Thus, this study considers the historic urban landscape, all natural and cultural attributes placed in historic layers of cultural accumulation, including hydrology and other natural characteristics of the place. In this way, the water presence in historic cities as a natural resource that circulates in rivers, fountains, underground flows, aqueducts, cisterns, and other cultural elements are very important to explore as living testimonies of the human presence and as a natural resource to preserve for the future. In fact, the water landscape and the water heritage have a social projection as a legacy and expression of the history of a certain place and the local communities. The charts and conventions written around the heritage conservation and preservation do not reflect around the water as a tangible or intangible heritage resource and do not debate the methodologies and the practical implications of the water use. It is a gap in the historical preservation and heritage conservation. Thus, this chapter proposes to highlight the role of the water heritage conservation as a way to develop cultural and sustainable tourism practices. The determination of principles around safeguard of resilient water landscape will improve the information and valorization around this enormous heritage. It is important to debate principles and a methodology that tourism and local communities could adopt and implement to protect water as a natural and cultural resource in a risk.

1.1 INTRODUCTION

Urban landscapes are the most heterogeneous. In fact, guarantying the sustainability of those landscapes can be challenging. In this context, Wu (2010) emphasized the importance of research in urban landscape sustainability. This issue must be taken seriously. Urbanized areas account for around 60% of residential water consumption will be taken into consideration (Grimm et al., 2008).

One way toward sustainability of urban landscape is conservation of its water resources, since it is a critical resource in operation and development of any city. In this context, furthermore, water is a natural and cultural, material and immaterial resource; at the same time is an aesthetic, social, political, environmental, and economic element in the cities (Freitas et al., 2017b).

Water resources are becoming scarce. Thus, there is an increasing demand to conserve water in urban landscapes. It should be realized that the sustainable use and protection of water resources in urban landscapes is more than the preservation of natural resources, it is also the preservation of area's cultural heritage. Concerns about the water heritage inventory arise following the UNESCO (2011) recommendations for the historic urban landscape that suggests the promotion of inventories of cultural and natural heritage in historic cities as a way to preserve and manage the urban landscape. History, heritage, and culture are an inspiration for our life. Furthermore, cultural heritage in historical cities is essential for appealing tourists to a destination.

However, nowadays, water landscape and water heritage are in a risk due to the various factors such as, climate alteration, change in land use, increasing demand for water resources, among others. One of the natural factors is climate change, as an accelerator process, that increases stress conditions in the landscape vulnerability, which could generate losses in tourist destinations. This continuous degradation of the natural and cultural heritage due to climate change can affect also the economic opportunities for local communities (UNESCO, 2016). In this context, the importance of preservation of water is considered as natural and cultural heritage in urban landscapes rises. UNESCO (2018) improved the defense of the nature–culture conservation strategy and developed, in partnership with the International Centre for the Study of the Preservation and Restoration of Cultural Property (ICCROM) and the International Union for Conservation of Nature (IUCN), the World Heritage Capacity Building Strategy to achieve innovation and excellence. Those institutions aim to use the capacity building as a methodology to a better cooperation between nature and culture in World Heritage places.

Following the definition exposed by ICOMOS (2013) capacity building "is the process, by which individuals, organizations, institutions, and societies develop abilities to perform functions, solve problems, and set and achieve objectives," The present chapter aims to contribute toward understanding the preservation of water as a natural and cultural heritage in urban landscapes, in this case, a World Heritage place. Capacity building within water as a cultural and a natural element can be an essential tool resulting in the positive overall performance of the stakeholders of a certain area, including individuals, organizations, institutions, and decision-makers.

As the case study area of this research has been selected the historical city Guimarães located in Portugal. GIS (Geographic Information Systems) tools have been used to map the landscape of the historical city center. Furthermore, the review of academic literature presented along with applied methodologies on water resources preservation in urban landscapes, in addition, the importance of capacity building experience is highlighted.

The specific objectives of the present study are (1) to map the historical landscape and water resources of the Guimarães city by using GIS tools. This mapping will identify the cultural heritage and the presence of water resources in this city classified as World Heritage. This mapping will improve the attention dedicated to water and promote the awareness of locals and visitors about the water present in a city, in a way of sustainable capacity building. (2) Debate the attained results, following with the most important theoretical contributions, practical implications, and suggestions for further research in the field of the cultural heritage linked with water that could prevent and theorize capacities in water conservation.

As UNESCO (2018) reinforces, a World Heritage place should guarantee the innovation and the excellence in the preservation and conservation of natural and cultural heritage and traditional techniques associated more sustainable then the actual ones.

1.2 WATER AS A NATURAL/CULTURAL RESOURCE: THE LITERATURE REVIEW

Water is vital for our life and for the well-being and survival of future generations. Water is the key element of the socio-economic development of the countries and is one of the most critical resources for the tourism industry (UNWTO, 2013). Nowadays, water scarcity and water quality challenges have been identified in many parts of the world (Kasim et al., 2014). Thus, it has been recognized that water is a natural resource, which needs to be preserved. In this context, water issues in the tourism industry are obtaining big attention from organizations, such as the United Nations Environmental Programme, the United Nations World Tourism Organization (Gössling et al., 2015).

Moreover, this resource is an unquestionable element of the urban planning. Historically the decision makers assumed its presence as a fundamental in city or rural planning. In the past, the water was central in urban places. All cultural, political, economic, and social life developed around water and its supply (Freitas et al., 2017a). Nowadays, the impressive growth of the population in urban areas and the implementation of water supply network, develop another relation with the water, in particular disappears the feeling of nature-dependent swallowing resource.

It is important to guarantee the sustainability of the cities by preserving its water landscape. Furthermore, the conservation of water landscape in historical sites is certainly advantageous for developing tourism since the tourism industry highly depends on the availability and quality of water resources. Moreover, this issue is critical for sustainable tourism management strategies and programs, which are actual nowadays. Since the tourism, as well as the climate change, could be a factor of rapid change in water culture and heritage not only in the process of the water use but also because of the water loss due to the growth of tourism and its impacts. Given this context, Kjelgren et al. (2000) expressed their concern regarding the conservation of water resources to support the population in urban landscapes, since the majority of the population is concentrated in urban areas.

Nowadays, the growth of the urban population has changed the balance between water demand and supply, which in turn make additional pressure on the environment. One of the most critical environmental problems in historical cities is water supply issues due to the growth of the urban population (Chang, 1998). The growth of the urban population leads to different water supply and water demand issues (Hilaire et al., 2008). The factors such as landscape irrigation, climate change, the decline of water quality, and public and private water consumption can influence supply and demand, that have big impacts on the water as a natural resource, in particular in urban places (Renouf et al., 2018).

Agreeing with Junqing et al. (2011), "water cultural heritages protection and sustainable utilization are linked together and form a continuous-unity. The development must submit to protection, and the first principle is to coordinate development with protection." That is where sustainable use of heritage resides.

Today the urban landscape is an accumulation of layers of past interventions of the man in a certain place (Maniglios, 2017). This accumulation of cultural layers of social expressions in a diachronic perspective, are historical documents of the societies living and inheritance. In this context, water as a natural and cultural resource, material and immaterial, has been an aesthetic, social, political, environmental, and economic element in the cities (Freitas et al., 2017b).

The concern about the water heritage inventory arises following the UNESCO (2011) recommendations for the historic urban landscape that suggests the promotion of inventories of cultural and natural heritage in historic cities as a way to preserve and manage the urban landscape. This document considers as historic urban landscape all natural and cultural attributes placed in historic layers of cultural accumulation, including hydrology and other natural characteristics of the place. In this way, the water presence in historic cities as a natural resource that circulates in rivers, fountains, underground flows, aqueducts, cisterns, and other cultural elements are very important to explore as living testimonies of the human presence in the city history and as a natural resource to preserve for the future.

The water landscape is full of resilient cultural elements (tangible or intangible) related to water. History, heritage, and culture are an inspiration for our life. Tuohino and Pitkänen (2004), in the case of thousand Finnish Lakes, discussed the symbolic importance of the lake's water for the Finnish culture. Wijetunga and Sung (2015) defined it as being "known as the combined work of humans and nature [...] can be defined as forms, relationships, and practices and are visible in any landscape with human involvement." The water has another advantage: it is an element that returns the man's action with emotions in the moments of leisure. The aesthetics, the social and cultural water behalf, and the "acoustic comfort is an important element of the landscape experience" (Ren and Kang, 2015). In urban places where the traffic and other noise are more intense, the water sounds are important.

However, nowadays, water landscape and heritage are in a risk due to the climate change as an accelerator process, that enlarges stress conditions in the landscape vulnerability, generating loses in touristic destinations. This continuous degradation of the natural and cultural heritage due to the climate change will affect also the economic opportunities for local communities (UNESCO, 2016).

Over the tangible heritage lie the threat of climate change and increasing of extreme event and change of the weather patterns, in particular, the increase of temperature and the decrease of humidity, the effect of strong winds that affect negatively and compromise the cultural heritage assets, the sea-level rising, and other natural occurrences as storms, floods, earthquakes, fire, and others (Sesana et al., 2018).

The urban development without precaution is another risk in consideration and the urban growing endanger the waterscape and water as natural and cultural heritage (Mitra and Banerji, 2018). In urban places, the identification of the cultural heritage related to water is important considering that a part of the water and its cultural and historic elements are underground and the land use without precautions of safeguard in urban places could damage or destroy these cultural elements or the natural course of the water causing negative impacts.

This study considers the historic urban landscape, all natural and cultural attributes placed in historic layers of cultural accumulation, including hydrology and other natural characteristics of the place. In this way, the water presence in historic cities as a natural resource that circulates in rivers, fountains, underground flows, aqueducts, cisterns, and other cultural elements are very important to explore as living testimonies of the human presence and as a natural resource to preserve for the future. In fact, the water landscape and the water heritage have a social projection as a legacy and expression of the history of a certain place and the local communities.

The historical analysis reflects the importance of the water and a historical landscape to preserve not only because of its natural importance but also because of its important cultural testimony of the passage of times and construction of the memories of the places. In this way, it is crucial to create a chart of principles around water and its uses and it is fundamental to discuss how the capacity building could expand the conservation of the water resilience and heritage and how to develop attitudes (from residents or from visitors) toward preserving water.

It is undeniable that cultural heritage, tangible or intangible, must be protected for its significance, value, and for the capacity to be a unique testimony of diversity, identity, and history on the planet Earth. As life testimony of human culture and human existence, in macro or micro places, cultural heritage needs to be preserved for future generations as irreplaceable elements of human construction through times.

Charts and conventions written around the heritage conservation and preservation do not reflect around the water as a tangible or intangible heritage resource and do not debate the methodologies and the practical implications of the water use. It is a gap in the historical preservation and heritage conservation to think water in its natural field and improve good practices to manage and conserve water in the territory, in particular in the stressed urban places.

1.3 CAPACITY BUILDING IN WATER MANAGEMENT AND CONSERVATION

Capacity building is getting bigger attention as a tool to improve the overall performance of many institutions. The expression "capacity building" is a constructive process of acquiring capacities, abilities, or skills. Alaerts (1996, p. 59) defined "capacity" as the "knowledge, skills, attitudes, and values as we find them in individuals, and as they are aggregated in organizations, communities, and in all other forms of arrangements that define individual and collective behavior." The capacities are an important part of any institution (organization) and play a key role in defining them. ICOMOS (2013) defined capacity building as a constructive process of individuals, organizations, institutions, and societies of acquiring and developing capacities to solve problems and reach the goals. In both definitions, it is possible to identify some keywords that explain objectively the concept: constructive process, collective process, capacity developing to achieve objectives.

Some researchers had various critical questions related to capacity building and suggested to link capacity-building priorities with water scarcity concept, arguing that it will be more effective in addressing water issues (Wolfe and Brooks, 2003). This issue is becoming crucial if we take into consideration global water consumption increased six times in the last century. Moreover, nowadays water scarcity is a major problem for over 40% of the world's population. It is estimated that if present water consumption patterns continue, global water demand will increase by 55% by 2050 due to the growing demand of various industries and domestic uses (Connor, 2015).

In addition, external factors such as climate change, increasing human impacts on the environment will demand the modern skills of the managers to address all those issues. This is the case of the protected areas, where various external pressures on the environment and growing demands from public sides are becoming a main issue in management (Teel et al., 2013). The authors discussed the results of the global review prepared for the 1992 World Congress on National Parks and Protected Areas and expressed their concern on poor management and policy being responsible for the threats rather than external factors in the protected areas. An inadequate legislation, poor management, a lack of staff, and skills demanding modern time have led to the threats in the protected areas. Thus, skilled human resources are needed to preserve the natural resources, among them water, which is crucial today. For example, over the past century economic achievement of South Africa was partially due to the competences and skills of its water managers. However, these achievements were distributed unevenly and most of the nation suffered from an inadequate supply of water resources. Therefore, water legislations and strategies were designed to try to address those issues (Colvin et al., 2008). Teel et al. (2013) also argued that multi-institutional partnership can enhance the leadership skills and abilities and technical competences of the managers of the protected areas in the case of India. Following this idea, special attention has to be made for the development of the capacities of the individuals along with capacity building among the institutions.

The importance of capacity building in water management has attracted the attention of the workers in this sector and researchers since the early 1990s (Franks, 1999; Timmer et al., 2007). For example, there were held two conferences on the capacity building process in the water sector in Delft (Alaerts et al., 1996). The first conference was held in 1991, and was dedicated more to explore the concept of capacity building in the water sector, while the second one was organized in 1996 and was concerned with the practical aspects of capacity building, as well as measures and activities which can be appropriate in a practice.

As a result of the first conference the following conclusion on the capacity building concept has been drawn: the capacity building consists of three elements, related to the creation of a favorable environment with appropriate policy and legal frameworks; development of human resource and improvement of managerial systems; as well as institutional development, including involvement of the community.

The second conference draw the practical results such as, decentralization and privatization policies, regulation of water prices and tariffs, improvements of the administrative procedures, development of human resources through education, training, networking, equipment of the libraries, etc. (Franks, 1999). As an outcome of the conference, human resources development, specifically improvement of skills, abilities, and competences of water managers have been considered as essential component of the capacity building. In this context, some literature mentioned the role of environmental education in the process of capacity building (Echevarría et al., 2013). However at the same time, we should keep in mind that focusing on only in development of human resources could be insufficient toward capacity building issues, also the development infrastructure for waters sector is essential. This is the case of historical areas, where its water landscape and cultural heritage are of high importance. Moreover, taken into consideration the increasing interest in cultural heritage among the tourists.

Growth of tourism and increasing pressures from urbanization in historical sites have created the need for an integrated approach to address the interactions among human activities and respect water resources of the area. Furthermore, historical sites have been widely recognized as essential economic resources in many countries. In this context, special attention is given to the cities, as the points favorable for infrastructure development, services, etc. This indicates the conservation of the urban landscape from cultural and historical perspectives. However, it should be noted that conservation of urban heritage and its water resources is not only important for its historical value but also for its potential to developing cultural and heritage tourism, which is one of the fast-increasing industries today. For example, World Bank's projects aim to support the conservation and management of cultural heritage assets (Ebbe, 2009). Therefore, the importance of preservation of water resources in historical places through capacity building process increases. Since an essential role of capacity building in water resources conservation, in general, have been already recognized.

Analyzing previous literature, we can observe that the importance of capacity building experience for coastal managers also has been emphasized (Echevarría et al., 2013), while others discussed its role for the managers of protected areas (Teel et al., 2013). Some authors discussed the application of capacity building in drinking water protection (Timmer et al., 2007).

A growing interest in application capacity building in water management can be explained by the increasing pressure on the environment and water resources. This is the case of Uruguay, where the development of various activities such as tourism, agriculture, transportation as well as the growth of population has made additional pressure on coastal zones. Unfortunately, not all consumers fully understand the priceless value of the coastal environment and the needs of its preservation (Echevarría et al., 2013). Therefore, in the integrated coastal zone management approach the key role of capacity building in improving the skills and abilities of the planners and managers in guaranteeing sustainable management of water has been highlighted.

However, at the same time, in the case of the water sector, it is also true that this sector has a huge coverage, starting simply from the provision of drinking water for humanity, water supply for food and irrigation (Franks, 1999), as well as being natural and cultural heritage of any nation. Thus, water managers can face certain challenges in management issues due to the needs to understand natural, ecological, social, and cultural factors.

Based on above-mentioned discussions, we can conclude that the role of capacity building in water resources preservation in historical sites is essential. First, the key players in the capacity building process of the water sector should be defined. Then, improving skills and competences in better and effective management should be of interest to the managers, planners, and decision-makers, while raising the knowledge in the protection of water resources should be applied for the local people. Moreover, the supportive policies and favorable environment should provide opportunities to apply capabilities, skills, and competences of waters sector workers.

1.4 THE HISTORICAL CITY OF GUIMARÃES

During the last decade, tourism in Portugal has experienced remarkable evolution. The impressive growth of the tourism industry in Portugal can be largely accomplished at the great expense of environmental quality. One of the rich heritage sites of Portugal is the historical city of Guimarães. This town is an ancient city of Portugal that inherited a rich traditional urban heritage for many years and should effectively use its historical and cultural advantage to enhance its image. The number of visitors to this historical site is increasing every year. Therefore, it is a good time to improve and restore the cultural and natural landscapes of its ancient city such as Guimarães, particularly its water landscape. Therefore, it should be of interest for the authorities and local people to take positive steps toward this process.

Guimarães is a city of Northern Portugal, implanted in the Region of Minho, genesis of nationality and with a strong symbolic significance for the country (Fig. 1.1). The tradition says that the first king of Portugal, D. Afonso Henriques was born in Guimarães. It belongs to a territory dominated by the hydrographic region of the river Cávado, in the North, and hydrographic influence of the river Leça, in the South, which control an area of 3584 km^2 (Soares, 2016). Within this area, highlight the hydrographic basin of the river Ave and the river Vizela, which delimit the county of Guimarães and it will be more relevant to our analysis.

The Ave basin covers an area of 1391 km^2, and is located in the zone of greater density population in the region. The Ave River is the largest watercourse in this area, and is born in the Cabreira Mountain at 1200 m of altitude. From here to its estuary in the Atlantic Ocean, crosses 101 km through a rugged and fertile territory. Its main tributaries are the river Este, on the right bank and the Vizela River, on the left bank (Soares, 2016). The whole region is rich in watercourses that fill a capillary web that extends throughout the region of Minho. It also has the highest precipitation levels in the country.

It is therefore a city with historical, cultural, and social prominence; over the time, that represents a remarkable synthesis of the national identity and the values of authenticity associated with it. It belongs to the district of Braga and its territory is included in the NUT III Ave sub-region.

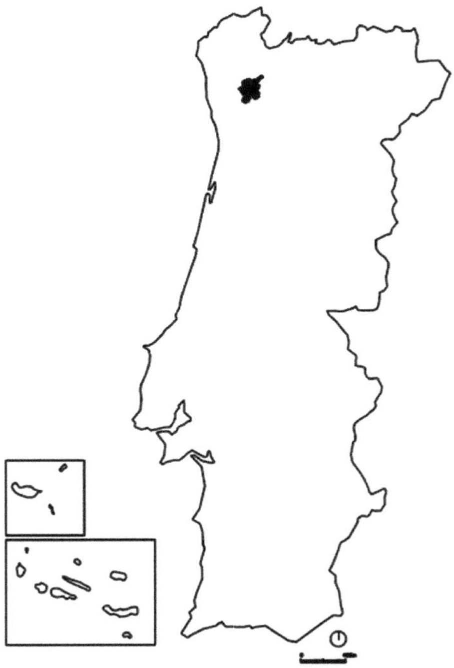

FIGURE 1.1 Map with Guimarães location.
Source: Authors' own compilation.

The Guimarães notoriety began in the Medieval Age, more concretely in the territorial reorganization of the Portucalense County (10th–11th century) and the future territory of Portugal (12th century). In the first period, the presence and influence of Countess Mumadona Dias, the aunt of King D. Ramiro II of Leon and widow of Count Hermenegildo Gonçalves, a tutelary figure of the city, were noted and influenced by both; the consolidation of the territory, or in the design of the medieval city. On the one hand, she led the creation of a convent to be built in the village of *Vimaranes*, on the other; she supports a castle, between 959 and 968, to defend this territory, to be erected on an adjacent hill (Ferreira, 2010). These two fundamental axes will be, henceforth, the poles of fixation of the people and the later development of Guimarães. From these crucial poles, the village will develop around *Vila Baixa* (where the convent and the settlement were located), and *Vila Alta*, where the military power was preserved symbolized by the tutelary presence of the castle. Both will remain, with numerous changes produced during 1000 years of history, in the urban landscape of the city, having persisted until the present day (Ferreira, 2010).

Later, in the second half of the 11th century, in pursuit of a strategy and consolidation the territory of the Portucalense county, Alfonso VI of Leon and Castile, grants to a knight of the Burgundy, the count D. Henrique, the government of the Portucalense Province has, for the purpose, married his daughter D. Teresa with this knight. From this union, was born in 1111, D. Afonso Henriques who would become the first Portuguese monarch (Mattoso, 1985). It is significant the importance of Guimarães in the territory of Portucale and the primacy it held among the other villages. This primacy will be accentuated during the early years of D. Afonso Henriques when he intends to support the independence of the Portucalense county, against the claiming power of Leon and Castile that wanted to force the medieval legal/political principle of submission that was expressed at the time, through the rite of homage and vassalage of the local nobility to the king (Mattoso, 1985).

With the evolution of the city and the new defense requirements, the two villages—*Alta* and *Baixa*—were merged through a walled enclosure that incorporated the two urban settlements. This intramural nucleus will melt the two villages, becoming one agglomerate, and will remain, with few urban changes after the 15th century, nowadays.

It is this intramural complex that will be recognized by UNESCO in 2001 as World Heritage due to the historical, social, and patrimonial value that it preserves and that allows an integrated understanding from the Middle Ages to the present day (Fig. 1.2).

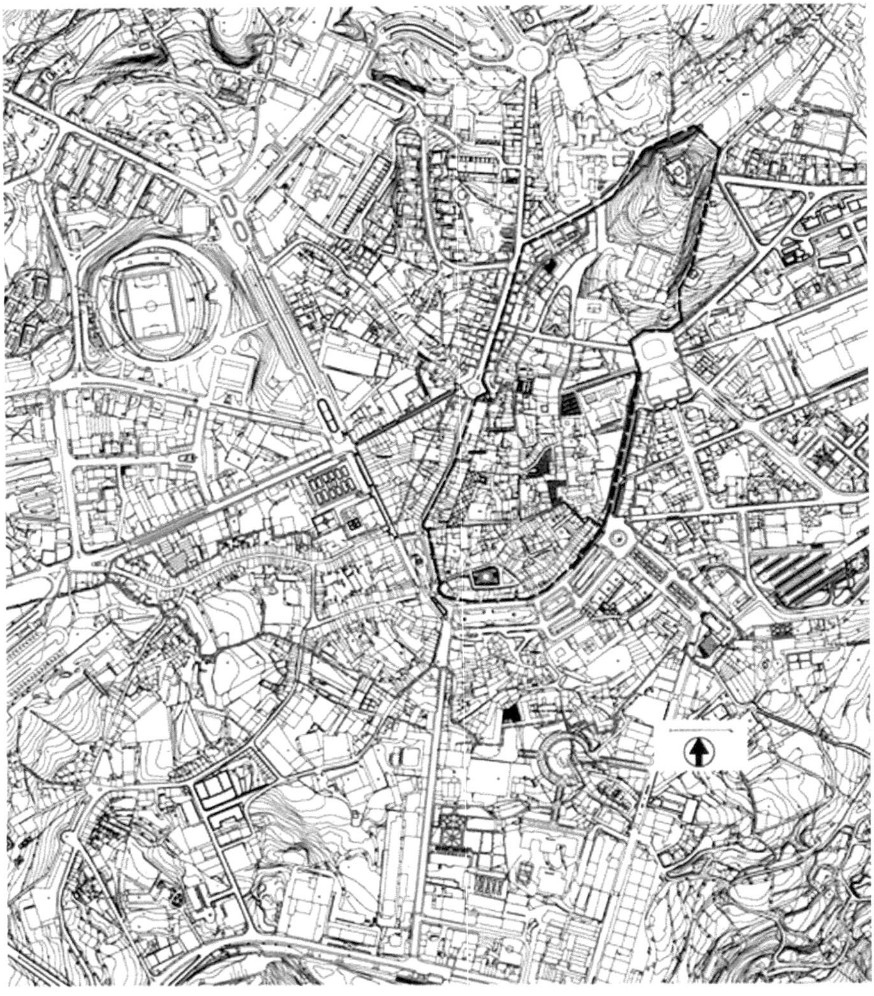

FIGURE 1.2 Areas of protection and cultural heritage.
Source: Authors' own compilation.

1.5 METHODOLOGY

The determination of principles around safeguard of resilient water landscape will improve the information and valorization around this enormous heritage. It is important to debate principles and a methodology that tourism and local communities could adopt and implement to protect water as a natural and cultural resource at risk.

The concern about the water heritage arises following the UNESCO (2011) recommendations for the historic urban landscape, that suggests the promotion of inventories of cultural and natural heritage in historic cities as a way to preserve and manage the urban landscape (Maniglios, 2017). Both documents insist on the landscape preservation, but do not refer water as the main natural and cultural resource that needs to take part of those debates. Those inventories are also important as a first step to validate the negative impacts, the causes, and promote strategic plans or decisions to keep way negative impacts. "The monitoring of hazardous events through change detection has an important role in emergency management. Such actions can be performed shortly after the hazardous event for first rapid mapping but also over longer periods of time for recovery purposes and risk mapping" (Cuca and Barazzetti, 2018).

Mapping and digitalization is the best way to preserve heritage (Di Stefano et al., 2016). Mapping the water is an unquestionable step for water conservation, especially in urban areas. "Mapping is a valuable tool in conservation efforts, as it may contribute to establish suitable heritage protection areas" (Martínez-Santos and Martínez-Alfaro, 2014). To map heritage, it is a way to promote knowledge about a thematic, in this case, the presence of the water in Guimarães; and debate the possibility to the contemporary conservation and urban planning strategies to enrich, in this case, the heritage inside a city classified by UNESCO as World Heritage.

In fact, water is a cultural, natural, aesthetic, social, political, environmental element in urban places in constant risk related to climate change, the historical process of natural and cultural passage of time, and other human actions as disruptive urban development. The methodological proposal to improve the conservation and preservation of the water and the water heritage in urban places is inherent in the use of mapping methodology to achieve the heritage valorization and interpretation in a process of capacity building of all actors in urban places. Mapping associated with the evaluative and interpretative process that the visual use of water in Guimarães history and capacity building is an important way to promote water conservation planning and management (Fig. 1.3). Following this model:

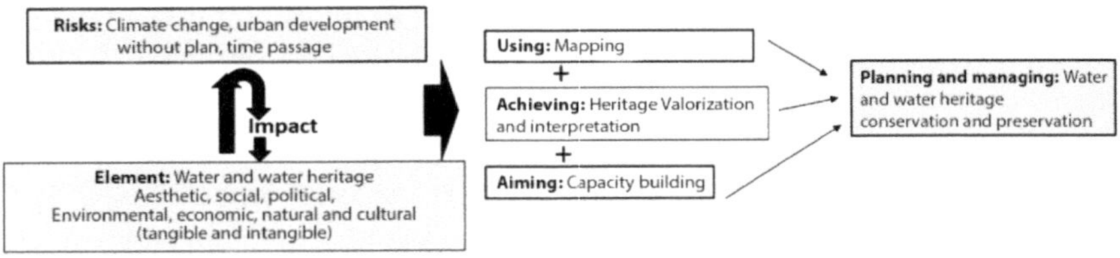

FIGURE 1.3 Conceptual framework.
Source: Authors' own compilation.

In this study, the option was to map the fountains in use in three different periods of the history of Guimarães:

1. The presence of the water in 15th Century, following the information of a historical map designed in 1569; Guimarães map of 1569—National Library Foundation—Brazil. This map that portrays the city in the middle of the 16th century also refers to some points of water supply distributed around the urban perimeter. Although we have access to only nine sources of water, expressly mentioned on the map; through other sources we can conclude that there are some others, inside and outside the wall perimeter.
2. The presence of water in the city in 1706, following the historical work made by António Carvalho da Costa, "Corografia Portugueza e Descripçam Topográfica do Famoso Reyno de Portugal," published in 1706, allows by the description of a travel itinerary denominated: *The Sources, that Villa of Guimarães has inside its walls, and in its Suburbs*, to know the sources, tanks, and wells that existed in the beginning of the 18th century in Guimarães (da Costa, 1706). It is an interesting picture that, in addition to identifying the places where the sources and wells were located, also gives us references on the water quality of the water fountains.
3. The 19th and 20th century was analyzed through the map of 1863 (plant of the city of Guimarães, Manoeld' Almeida Ribeiro, 1863),[1] in combination with the articles published by João Gomes de

[1]Access in https://archeevo.amap.pt/details?id=131117 (accessed Dec 21, 2018).

Oliveira Guimarães, the Abbot of Tagilde, in Revista de Guimarães, referring the water supply in Guimarães in *Apontamentos para a História do Concelho de Guimarães*, from the beginning of the 20th century.

1.6 MAPPING THE WATER AND WATER HERITAGE IN GUIMARÃES

The water supply system in Guimarães does not differ from other similar cities. Its conduction was made through a network of pipes, chapels, or aqueducts that originated in springs, wells, or mines that were used to provide the necessary flow to the population of the village (Fig. 1.4). The main source was located, since medieval times, in the Penha Mountain, very near to the city. These water supply networks invariably ended up in fountains or tanks distributed throughout the urban perimeter (Veloso, 2012). However, unlike other cities in the country, such as Porto, Lisbon, or Coimbra, Guimarães does not have a large river, where it can supply its needs. The city has some brooks, such as the Couros or the Herdeiro and the Selho, that constitute the hydrographic network of the city. Based on this amount of information, we can come to some conclusions about the distribution of water in the city and its surroundings.

By the analysis of the map of 1569 it is possible to conclude that the number of public fountains within the walls is reduced. We have pointed out only the fountain of the Oliveiras square (31), in the neuralgic center of the village, attached to the tower of the *Colegiada* church. The map (Fig. 1.5) is silent as regards other water supply points within the urban perimeter. However, we can mention the existence of other points of supply, such as the Well of the Door of Senhora da Graça or Santa Luzia (35), the Well of the Square of the Peixe (36), Fountain of the house of the Arc (38), this one private, the remains of the sources of the Convent of Santa Clara (37), recently founded in 1553, compared to the date of the drawing of the map, in 1569. A reference is also made in the 1706 travel itinerary, which reveals existence of numerous private wells in the intramural houses, in their backyards, where there are one, two, three and sometimes four wells, which served to supply their owners, something that we can conclude, with a good margin of safety, as a much older reality (Freitas and Leite, 2018).

FIGURE 1.4 Guimarães hydrographic map.
Source: Authors' own compilation.

The fountains of Guimarães are also located outside of walls. They are mentioned in the map of 1569 fountains that were mostly located along roads or trails (Estrela, 2017), as well as the construction of the monasteries of S. Francisco and S. Domingos in the 13th century, firstly modest constructions near the wall and which were therefore demolished by order of King D. Dinis, later displaced to the place they occupy today, the surrounding areas underwent a remarkable urban development. The supply of potable water was an incentive so it will not be surprising, the profusion of sources throughout this area coupled with the fact that there is a need for constant water sources for the leather industry through the stream that made its way there. We have indicated in the map of 1569 (Fig. 1.5), the Fountain of the Abade (01), Fountain of the Duchess (05), Fountain of the Americas (12), Fountain of the Pipa (07) or the Tank of S. Torcato (14), these next to communication ways, and fountains and tanks in the area of influence of the mendicant monasteries, such as Fontain da Quintã (15) or Fontain do Banhadoiro (03).

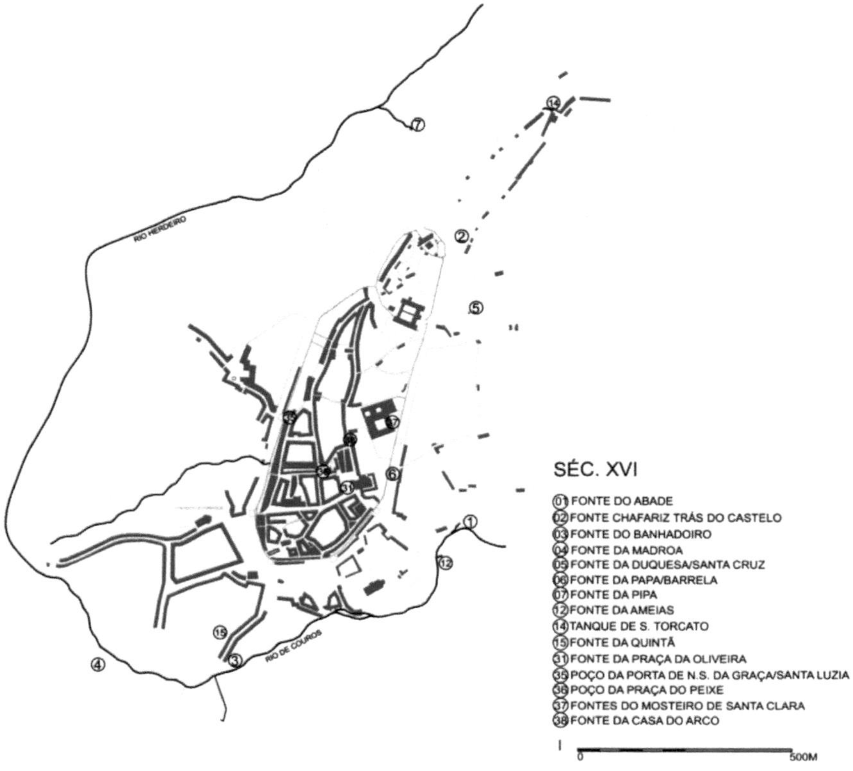

FIGURE 1.5 Map of the fountains in 16th century.
Source: Authors' own compilation.

The *Corografia* of Father António Carvalho da Costa, published in 1706, allows the description of a travel itinerary which he calls *Fontes, que a Villa de Guimaraenstem dentro dos seusmuros, e nosseus Arrabaldes*, to know the fountains, tanks, and wells that existed at the beginning of the 18th century in Guimarães (Fig. 1.6). It is an interesting picture that besides the identification of the places where the fountains and wells were located, gives us also references, on the water quality of the springs. The combination of these sources, described in *Corografia*, with those of the map of 1569, allows to verify that many of the named ones already existed previously, adding others for which we do not have references that permit assumptions about its existence in the middle of the 16th century. We have ascertained the sources 01, 02, 03, 04, 05, 06, 07, 12, 14, 15, 31, 35, 36, 37 and 38, as already existing in the middle of the century, and which were active in 1706, and others marked as 16, 10, 09, 18, 30, 40, 45, 41, 42, 43, 44, 23 which are the result of incorporations between 1569 and 1706. In this period there is a transformation in the planning of these structures. Fountains and elaborated water fountains begin to emerge, which are constructive elements that are worthy by their own significance, that is, they are independent of architectural supports and accentuate this symbolic character that had already been acknowledged before. They are in the center of squares and *rossios*, consolidating the space and the power of representation; "fountains and elaborated water fountains appear isolated in the center of the squares and, as a modern element, possibly arising from the 15th century, amplified by aesthetic principles" (Estrela, 2017). In the case of Guimarães, within this period, stands the Toural square fountain built in 1588 by Gonçalo Lopes (marked on the map with number 23).

Concerns about the control, management, and care of public water supply points in urban centers, as demonstrated by the authorities have been well documented since the dawn of the Middle Ages in Portugal and are becoming more pronounced as we move forward in time. Disputes arising from the quarrel over water sources, hygiene, abuses or control of mines, and springs are frequent in many cities, as Maria Teresa Nobre Veloso shows in the city of Coimbra since the 11th century (Veloso, 2012), and tends to intensify with growth and affirmation of urban development. In Guimarães, during this period, between 1569 and 1706, there are innumerable cases that can be removed from municipal documentation and from litigation sponsored by other powers, such as religious, which in the Guimarães case is indelibly marked by the omnipresence of the Royal Colegiada. The management of water resources, outmost as a result of population growth and increasingly specialized urban organization start to have crucial importance at both the public and private levels.

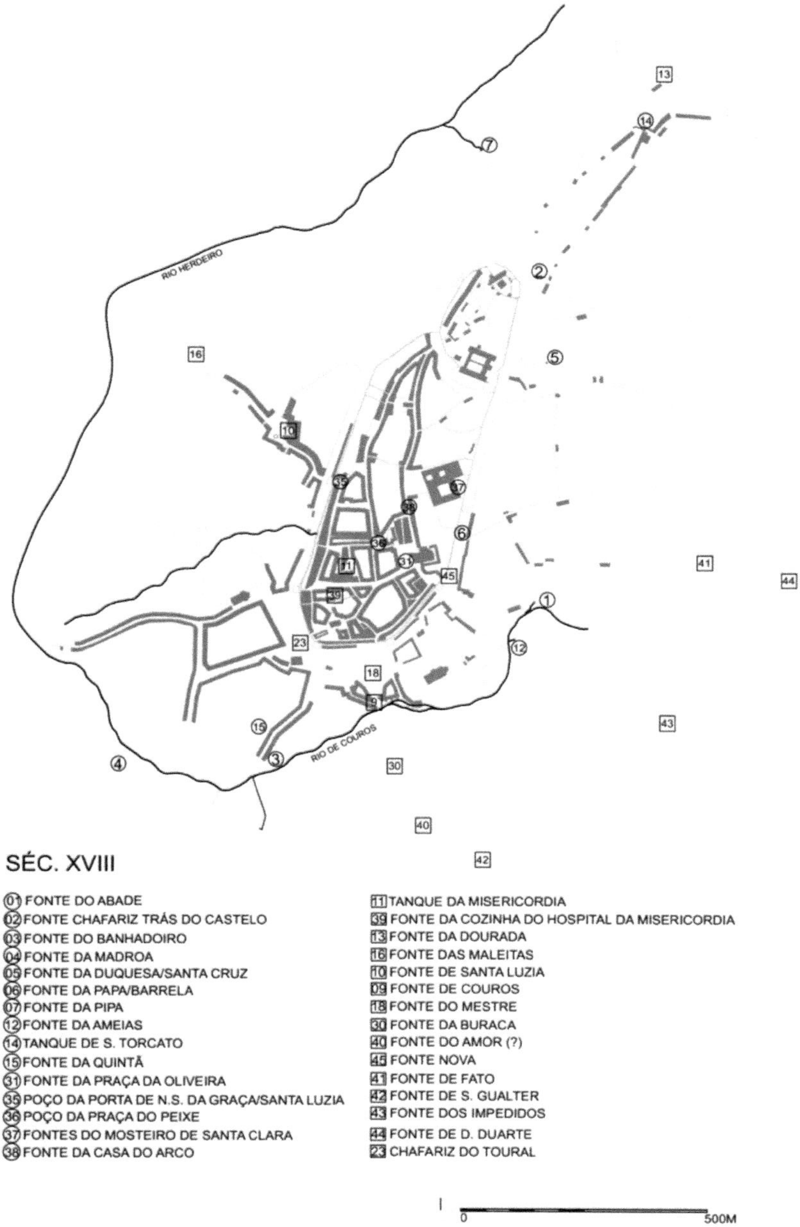

FIGURE 1.6 Map of the fountains in 18th century.
Source: Authors' own compilation.

In the mid-19th century, concern about water supply to urban centers remains a constant reality. The need to face the increase consumption, among other factors, due to demographic pressure, regularization of flows throughout the year, as well as concerns about the quality of water for human consumption, in the light of new advances in public hygiene, the awareness of urgency in the municipal policies of water supply in urban environments (Fig. 1.7). Guimarães, due to a chronic insufficiency in the supply flows, especially in the summer, very pronounced from the mid-80s and reaching its peak in 1898, in which there are reports of fountains and sources that dried, stimulate, from the end of the 19th century the need of more attention to this theme. Since the middle of the 16th century, urban supply was mainly made by the water volumes in the Serra de Santa Catarina (as known as Serra da Penha), near the urban center, that were led through pipes to the city. These acquisitions of new mines and their reservoirs were acquired by the public authorities over time, in the 18th and 19th centuries, and introduced into the supply system. However, it was proved that it was not sufficient for the required levels of consumption; the constant deviations, the lack of maintenance of the structures, and the lack of cleaning that clogged the pipes and the water collectors produced a chronic lack of sources to the public supply. This led to a very marked demand for private solutions as wells that existed in large numbers for the city and its surroundings. The main issue, however, was the quality of the water for public consumption, either the one responsibility for the public authorities, through the network of aqueducts and fountains, or in the wells for private use. In fact, the existence of diseases caused by the consumption of contaminated water is a problem for the public authorities and hence the various causes of conflicts with the prohibition of certain fountains by contamination. The same was true of groundwater contaminated with sources of contamination and a lack of collective awareness about the problem of sewers and urban sanitation networks. This problem of lack of water resources for consumption and for industry became more accurate at the end of the 19th century. The first consolidated plans for obtaining sufficient flows for the definitive resolution of this chronic lack of water for public supply by the Swiss geologist Paul Choffat, as well as, in 1899, an in-depth study on the quality, physical, chemical, and bacteriological of the available sources, ordered to the microbiologist Charles Lepierre. The conclusion of these studies pointed, regarding supply, to increase the exploration in Serra da Penha, since long time the exceptional place for great volume of water in the city, and the quality of the water available for public consumption, the conclusion was the verdict of contamination of the flows in general (Costa, 2014).

It will be from these reports and from the awareness of the public powers, that at the beginning of the 20th century, will definitely be resolved the issue of quantity and quality in public water supply.

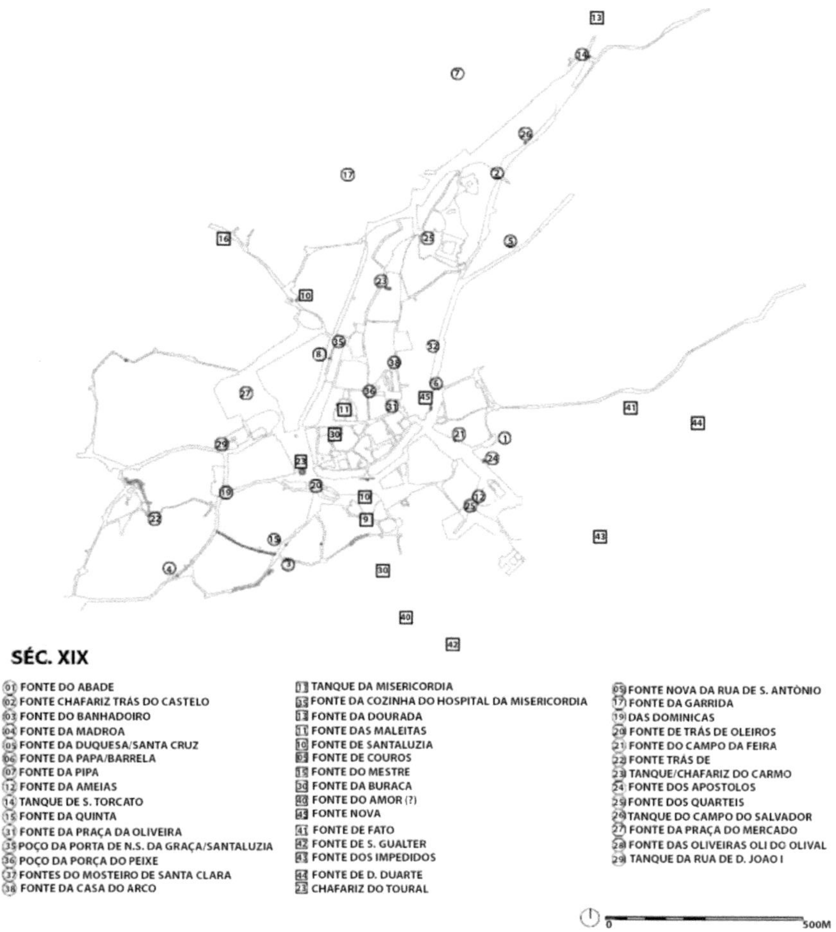

FIGURE 1.7 Map of the fountains in 19th century.
Source: Authors' own compilation.

1.7 CONCLUSIONS

Water is a cultural, natural, and aesthetic element of the historical city of Guimarães. Thus, maintaining its sustainability is of high interest for the policymakers, private organizations, and local people. It is even more important if we take into account that the aesthetic of the city also depends on good quality and availability in water resources.

The historical presence of water sources in the city refers to an urban hydraulic network in which the rivers Heir and Leather are predominant. The presence of wells and fountains scattered around the urban perimeter allowed to identify the points where the water in its natural state circulates through the city. The mapping is useful in monitoring process of water sources by taking into consideration other natural elements of the city (land using).

Based on the results of present work, we can see that over the time, the urban development has led to growth of number of water supply points accessible to the population. The urban growth has been accompanied with expanding water use from various perspectives up to the 19th century. However, it should be noted, that growing number of water sources lead to growing pressure on environmental state of water resources.

Obviously, first of all, water is used as a vital element in the daily lives of the population. At the same time, interest in the city as a historical heritage has led to a growing number of visitors to the area during the last decades. Consequently, the consumption of water resources has increased. Thus, the effective management of water has always been a problem that demanded the responsiveness from the public authorities, which is even more pressing when it comes to urban areas, which is characterized by its complexity. In addition, management of this fundamental good has always been a concern for the local population, having pressure from the legislation on this subject and at the same time searching for solutions. There are many examples of local authorities' intervention in regulating access to the exploitation of water, even when there are abundant resources. This was mostly related to social issues, such as arising disputes between local people and authorities on hygiene and protection against epidemics, the divine or religious properties of some sources or human contact of a playful nature, as well as in technological terms, with implications for urbanism, capture, conduction, quality, flow, and distribution (not forgetting the drainage of dirty water), constituting a constant concern, and over time, until the beginning of the 20th century (Gonçalves, 2017).

The above-mentioned concerns from the part of the local people and authorities are understandable, since the water is an element of easy contamination. Therefore, the concern about preserving this resource should be constant. The concern is accelerated by climate changes, increasing water consumption indicators, and pollution of water bodies worldwide. Thus, urgent measures are needed toward water conservation and protection.

One of the ways proposed in the current chapter is enhancing capacity building of the institutions and organizations involved in the water management process. In addition, it is suggested to local governments to involve community members in source water protection planning initiatives. This issue becomes more essential for the tourist destination such as historical city Guimarães. Since the image of the city's landscape highly depends on the state of its water resources as a natural and cultural heritage. Thus, appreciation of this resource is crucial for aesthetics, visual, and sensorial experiences of the visitors. Raising awareness among the visitors on the importance of protection of water would make them feel encouraged to contribute toward this process. This can reflect as another part of the capacity building process in the protection of water from both natural and cultural perspectives.

In general, we conclude that capacity building plays a significant role in the development, implementation, and maintenance of water resource protection strategies and programs. A previous research also matches with our suggestions on the role of capacity building in water preservation and management (Echevarría et al., 2013; Franks, 1999; Timmer et al., 2007).

While this work has been important in understanding water resources conservation and management in historical places through capacity building. Further research will be useful in developing the methodology on application capacity building techniques in water conservation process in historical places.

Over time in numerous urban centers the growing need for the urbanization of historic areas to the perception of new concepts of urban management, fundamentally in the 19th century, brought many problems to secular structures that caused dilemmas in the concepts of modernization. Water systems, aqueducts, fountains, or tanks were among the structures that were most damaged, not only because they were sometimes obstacles to progress but fundamentally because they represent an old-fashioned public system that is not hygienic and insufficient to the light of new forms of water supply for urban consumption, from now on effectively controlled by local authorities.

This led to the destruction of several notable specimens such as aqueducts or fountains, as well as to the permanent displacement of other structures that were being located at several places or dismantled.

Today, there is a remarkable tendency to recover this memory with the placement of some of these fountains, still reusable, either returning to their places of origin, as is the case of the Largo do Toural Fountain (23), or, if not possible, give them a dignified location, as is the case of the Fonte Nova St. António Street (08).

Of course, it is no longer the needs of public supply but the increase of policies to improve public spaces, through the need to create fresh locations and increase the humidity in decent values, as well as the connection with memory, that is, with collective assets. In this way, the past tends to unite again with the present, not in its primary function, public supply of water for consumption, but in improving the quality of public spaces.

KEYWORDS

- **water**
- **heritage**
- **conservation**
- **landscape**
- **tourism**

REFERENCES

Alaerts, G. J. Background Paper Capacity Building as Knowledge Management: Purpose, Definition and Instruments. In *Water Sector Capacity Building: Concepts and Instruments, Proceedings of the Second UNDP Symposium on Water Sector Capacity Building*. Delft, The Netherlands, 1996.

Chang, S. D. Beijing: Perspectives on Preservation, Environment, and Development. *Cities* **1998,** *15* (1), 13–25.

Colvin, J.; Ballim, F.; Chimbuya, S.; Everard, M.; Goss, J.; Klarenberg, G.; ... Weston, D. Building Capacity for Co-operative Governance as a Basis for Integrated Water Resource

Managing in the Inkomati and Mvoti Catchments, South Africa. *Water SA* **2008,** *34* (6), 681–689.

Connor, R. *The United Nations World Water Development Report 2015: Water for a Sustainable World*, vol. 1; UNESCO Publishing, 2015.

Costa, F. D. S. Mãe-d'água: Una contribución sobre el patrimonio hidráulico y el abastecimiento público de Guimarães (noroeste de Portugal). *Agua y Territorio* **2014,** *3*, 77–88.

Cuca, B.; Barazzetti, L. Damages from Extreme Flooding Events to Cultural Heritage and Landscapes: Water Component Estimation for Centa River (Albenga, Italy). *Adv. Geosci.* **2018,** *45*, 389–395.

da Costa, A. C. *Corografia Portugueza, E Descripçam Topografica Do Famoso Reyno De Portugal*, vol. 1; Deslandes, 1706.

Di Stefano, M.; Salonia, P.; Ventura, C. Mapping and Digitizing Heritage Sites: ROVINA Project for Programmed Conservation. *Procedia Soc. Behav. Sci.* **2016,** *223*, 944–951.

Ebbe, K. Infrastructure and Heritage Conservation: Opportunities for Urban Revitilization and Economic Development, 2009.

Echevarría, L.; Gómez, A.; Piriz, C.; Quintas, C.; Tejera, R.; Conde, D. Capacity Building for Local Coastal Managers: A Participatory Approach for Integrated Coastal and Marine Zones Management in Uruguay. *Revista de Gestão Costeira Integrada-J. Integr. Coast. Zone Manag.* **2013,** *13*(4), 445–456.

Estrela, G. F. Fontes e Chafarizes. O abastecimento de água nos espaços públicos na Baixa Idade Média portuguesa. Dissertação realizada no âmbito do Mestrado em Arqueologia// Faculdade de Letras da Universidade do Porto//setembro de, 2017.

Ferreira, M. D. C. F. *Guimarães, "duas vilas, um só povo": estudo de história urbana, 1250–1389*, 2010.

Franks, T. Capacity Building and Institutional Development: Reflections on Water. *Public Admin. Develop. Int. J. Manage. Res. Pract.* **1999,** *19* (1), 51–61.

Freitas, I., V.; Leite, P. P. Uma casa em Guimarães. Mª Isabel del Val Valdivieso. In *Expresiones del poder en la Edad Media*; Martín Cea, J. C.; de la Vega, D. C. Eds.; Valladolid: Universidad de Valladolid, 2018.

Freitas, I. V.; Martin Cea, J. C.; Del Val Valdivieso, I. Water Landscapes and History. *Revista Rosa dos Ventos—Turismo e Hospitalidade* **2017a,** *9* (3), 389–404. DOI: http://dx.doi.org/10.18226/21789061.v9i3p389.

Freitas, I.; Martín Cea, J. C.; Villanueva Zubizarreta, O.; Val Valdivieso, I. Abordagem metodológica de conservação da história e do patrimônio: O Vale do Douro. *Revista Tempo e Argumento* **2017b,** *9* (21), 99–125. Disponível no Repositório UPT, http://hdl.handle.net/11328/1928.

Gonçalves, L. R. Sistemas de circulação de água e poder na Lisboa medieval—séculos XIV a XVI. *Cadernos do Arquivo Municipal.* ISSN 2183–3176. 2ª Série N.º **2017,** 8 (julho–dezembro 2017).

Gössling, S. New Performance Indicators for Water Management in Tourism. *Tourism Management* **2015,** *46*, 233–244.

Grimm, N. B.; Faeth, S. H.; Golubiewski, N. E.; Redman, C. L.; Wu, J.; Bai, X.; Briggs, J. M. Global Change and the Ecology of Cities. *Science* **2008,** *319* (5864), 756–760.

Hilaire, R. S.; Arnold, M. A.; Wilkerson, D. C.; Devitt, D. A.; Hurd, B. H.; Lesikar, B. J.; ... Pittenger, D. R. Efficient Water Use in Residential Urban Landscapes. *Hort. Sci.* **2008,** *43* (7), 2081–2092.

ICOMOS. Principles for Capacity Building through Education and Training in Safeguarding and Integrated Conservation of Cultural Heritage, 2013. file:///C:/Users/ifreitas/Desktop/capacity%20building/ICOMOS_CIF_PrinciplesCapacity_EN_20130930.pdf (accessed Feb 20, 2019).

Junqing, H.; Xue, X.; Hong, Y. The Protection and Utilization of Water Cultural Heritages in Xi'an of China. In *2011 International Symposium on Water Resource and Environmental Protection*, vol. 4; IEEE, 2011; pp. 2594–2596.

Kasim, A.; Gursoy, D.; Okumus, F.; Wong, A. The Importance of Water Management in Hotels: A Framework for Sustainability through Innovation. *J. Sustain. Tour.* **2014,** *22* (7), 1090–1107.

Kjelgren, R.; Rupp, L.; Kilgren, D. Water Conservation in Urban Landscapes. *Hort. Sci.* **2000,** *35* (6), 1037–1040.

Maniglios, A. C. Chapter 3 Landscape and Education. In *Landscape Dimensions, Reflections and Proposals for the Implementation of the European Landscape Convention.* Council of Europe, 2017.

Mattoso, J. A história das paróquias em Portugal. *Portugal Medieval: Novas Interpretações* 1985, 37–56.

Mitra, D.; Banerji, S. Urbanisation and Changing Waterscapes: A Case Study of New Town, Kolkata, West Bengal, India. *Appl. Geogr.* **2018,** *97*, 109–118.

Martínez-Santos, P.; Martínez-Alfaro, Pedro E. A Priori Mapping of Historical Water-supply Galleries Based on Archive Records and Sparse Material Remains. An Application to the Amaniel qanat (Madrid, Spain). *J. Cult. Heritage* **2014,** *15*, 656–664.

Ren, X.; Kang, J. Effects of the Visual Landscape Factors of an Ecological Waterscape on Acoustic Comfort. *Appl. Acoust.* 2015, *96*, 171–179.

Renouf, M. A.; Kenway, S. J. Evaluation Approaches for Advancing Urban Water Goals. *J. Indus. Ecol.* 2017, *21* (4), 995–1009.

Renouf, M. A.; Kenway, S. J.; Lam, K. L.; Weber, T.; Roux, E.; Serrao-Neumann, S.; ... Morgan, E. A. Understanding Urban Water Performance at the City-region Scale Using an Urban Water Metabolism Evaluation Framework. *Water Res.* 2018, *137*, 395–406.

Sesana, E.; Gagnon, A. S.; Bertolin, C.; Hughes, J. Adapting Cultural Heritage to Climate Change Risks: Perspectives of Cultural Heritage Experts in Europe. *Geosciences* **2018,** *8*, 305; DOI: 10.3390/geosciences8080305.

Soares, E. Z. *Os Planos de Gestão de Região Hidrográfica em Portugal Continental: contributo para o desenvolvimento de um instrumento para a avaliação de planos de recursos hídricos.* Doctoral dissertation, Universidade do Minho, 2016.

Teel, T. L.; Carlos, D.; Andrew, W.; Manfredo, M. J.; Mathur, V. B. A Multi-institutional Partnership to Build Capacity for Effective Protected Area Management in India. *J. Park Recreat. Admin.* **2013,** *31* (2), 154–162.

Timmer, D. K.; De Loë, R. C.; Kreutzwiser, R. D. Source Water Protection in the Annapolis Valley, Nova Scotia: Lessons for Building Local Capacity. *Land Use Policy* **2007,** *24* (1), 187–198.

Tuohino, A.; Pitkänen, K. The Transformation of a Neutral Lake Landscape Into a Meaningful Experience–interpreting Tourist Photos. *J. Tour. Cult. Chang.* **2004,** *2* (2), 77–93.

UNESCO. Recommendation on the Historic Urban Landscape, 2011. http://portal.unesco.org/en/ev.php-URL_ID=48857&URL_DO=DO_TOPIC&URL_SECTION=201.html (accessed Apr 6, 2018).

UNESCO. World Heritage and Tourism in a Changing Climate, 2016. https://whc.unesco.org/en/tourism-climate-change/ (accessed Feb 20, 2019).

UNESCO. Convention Concerning the Protection of the World Cultural and Natural Heritage, 2018. file:///C:/Users/ifreitas/Desktop/capacity%20building/whc18-42com-6-en.pdf (accessed Oct 20, 2019).

Veloso, M. T. N. A água na cidade de Coimbra durante a Idade Média. *Revista Portuguesa de História*– t. **2012,** *XLIII*, 159–181.

Wijetunga, C. S.; Sung, J. S. Valuing the Cultural Landscapes Past and Present: Tea Plantations in Sri Lanka. *Landsc. Res.* **2015,** *40* (6), 668–683.

Wolfe, S.; Brooks, D. B. Water Scarcity: An Alternative View and Its Implications for Policy and Capacity Building. *Nat. Resour. Forum* **2003,** *27* (2), 99–107.

World Tourism Organization. *Tourism and Water. World Tourism Day*; UNWTO, 2013. http://cf.cdn.unwto.org/sites/all/files/pdf/unwto_annual_report_2013_0.pdf (accessed Feb 20, 2019).

Wu, J. Urban Sustainability: An Inevitable Goal of Landscape Research, 2010.

CHAPTER 2

CAPACITY BUILDING FOR SPIRITUAL HERITAGE TOURISM: THE CASE OF INDIA AND PAKISTAN

ANITA MEDHEKAR[1*] and FAROOQ HAQ[2]

[1]*Central Queensland University, Australia*

[2]*Canadian University Dubai, UAE*

[*]*Corresponding author. E-mail: A.Medhekar@cqu.edu.au*

ABSTRACT

Capacity building and raising awareness through education and training is key to conservation, development, and preservation of centuries' old spiritual heritage tourism sites, places, and spiritual places of worship, monuments, cultural, and natural heritage in India and Pakistan. India and Pakistan have numerous centuries old sites from the ancient to the medieval period; Buddhist, Islamic, and Hindu spiritual heritage tourism sites and monuments, which need to be preserved through capacity building and making people aware of conservation of these cultural and natural spiritual heritage. These spiritual cultural and natural heritages have to be protected and preserved from natural disaster and vandalism by the people and visitors for future generations. Public and private partnerships are required for capacity building at global, national, and regional level to reap the cultural, social, recreational, and economic benefits for sustainable long-term impact from spiritual heritage preservation, protection, and promotion.

2.1 INTRODUCTION

This chapter examines the capacity building for preservation, protection, promotion, and development of "spiritual heritage tourism" (SHT) of eight major religious heritage sites present in India and Pakistan. Heritage tourism is being appreciated and recognized by tourism agencies around the world in general, and by government agencies in particular. The low cost of product development, high natural differentiation, flexible delivery, rich spiritual and cultural heritage, and high tourist loyalty are the key features of heritage tourism (Haq and Medhekar, 2018). Heritage tourism is also known to augment the capacity building of a place or a nation based on education, training, conservation workshops, marketing, sustainability, site identity, and heritage preservation for the future generations by involving local communities.

Heritage tourism also helps to create innumerable opportunities for employment in heritage education, heritage management, preservation, conservation, building and architecture, heritage culture and arts, zoologist, botanist, and capacity-building training and workshops for heritage tourism awareness programs. Further, countries with population belonging to various religious minorities in particular, should include "SHT" education at school and university level to promote preservation, peace, all-inclusive unity-in-diversity of centuries old rich spiritual cultural heritage in India and Pakistan for the following eight religious sites: (1) Buddhism, (2) Christianity, (3) Hinduism, (4) Islam, (5) Jainism, (6) Judaism, (7) Sikhism, and (8) Zoroastrianism. Furthermore, it is essential to also include protection and conservation of the "spiritual nature of natural heritage," such as the Himalayan mountains, lakes, forests, and nature's beauty with embodies various spiritual heritage places of worships also needs capacity building programs for its preservation for the future generations.

Capacity building, therefore, is essential for these ST heritage monuments to be marketed as a vehicle of peace for social transformation and economic progress of the two countries, which have shared the history of spiritual heritage sites. Tourism management, government, and operators around the world are recognizing the growth of SHT amidst the proliferation of various types of fun-based tourism (Seraphim and Haq, 2018). For such a task one needs capacity building within the countries and between the countries in mutual partnership to build values of preserving and promoting such SHT for economic prosperity and peace between the two nations. However, SHT development and preservation is faced with many opportunities, benefits, and challenges in countries such as India and Pakistan. Therefore, it is important to present to the world the untapped heritage, in order to reap the social, cultural, recreational, and economic benefits from capacity building through SHT.

SHT education for preservation of spiritual heritage is the need of the century in many developing countries, which will be analyzed in this chapter with the purpose of capacity building to identify challenges to develop and preserve spiritual heritage (Bandyopadhyay et al., 2008; Haq and Medhekar, 2017; Timothy and Boyd, 2003) for employment opportunities, economics progress, and prosperity of the regions where these spiritual heritage sites are located. According to Simone-Charters and Boyd (2010), spiritual/religious/pilgrimage tourism is a segment of cultural heritage tourism. It is one of the oldest forms of travel since human history and most significant in context of volume of tourists for most religions of the world (Haq and Medhekar 2015; Sharpley and Sundaram, 2005; Shinde, 2007; Timothy and Olsen, 2006).

India and Pakistan are made up of people following many diverse religious spiritual orientations, languages, and cultures. This multi-faith diversity provides a competitive edge to be a world leader in the market of spiritual tourism as a vehicle for global peace (Haq and Medhekar, 2013). From historical times, followers of Hinduism, Islam, Sikhism, Buddhism, Jainism, Christianity, Judaism, and Zoroastrians have been living peacefully and have built tangible and intangible spiritual heritage sites for centuries. Both countries are blessed with built spiritual monuments, places of worships, and gardens linked to the major religious and spiritual beliefs, hence presenting the world with some of the best untapped religious centers and spiritual nature in the world (Chakraborty, 2011; Kumar and Singh, 2015; Medhekar and Haq, 2009; Shinde, 2007). However, what is required is to build capacity for awareness to develop and preserve SHT sites within the country, and between the two countries due to shared mutual history, culture, spirituality, and civilization for economic prosperity and peace.

The purpose of this chapter is to: (1) examine the significance of SHT development in India and Pakistan with focus on eight key religions. (2) Propose capacity-building strategies for SHT development and preservation for economic progress, prosperity, and peace of the region based on spiritual heritage attractions. (3) Evaluate the challenges and opportunities faced in development of SHT and benefits for the local economy. (4) Propose a public–private partnership (PPPs) framework that could be adopted in India and Pakistan for capacity building to develop, preserve, and promote SHT as a peace initiative and submit proposals for nomination to UNESCO world heritage listing. This chapter aims to provide a platform for capacity building with the aim for preservation and promotion of SHT for economic development in existing and neglected cultural and natural SHT in India and Pakistan, and empower the local communities for protection and preservation of the spiritual heritage sites for their sustainable future.

This chapter is structured as follows. The first introductory section of the chapter introduces the growing economic importance and significance of SHT development and preservation. Section 2.2 provides the literature review on SHT development, preservation, and promotion for socioeconomic progress and prosperity, in case of India and Pakistan. Section 2.3 discusses the case study examples from India and Pakistan SHT site development, preservation, and promotion for economic prosperity and nomination of the spiritual heritage sites for UNESCO world heritage listing. Section 2.4 outlines challenges and opportunities of capacity building through SHT development and preservation for nomination to UNESCO for economic prosperity and peace. Finally, Section 2.5 discusses and recommends PPPs needed for capacity building within and between the two nations to develop, preserve, and market SHT sites and managerial and policy implications, conclusion, and future research directions to advance the knowledge in the field of SHT as a vehicle to promote peace and economic prosperity and progress between the two nations.

2.2 LITERATURE REVIEW

Capacity Building: The International Centre for the Study of the Preservation and Restoration of Cultural Property is consulted for capacity building within world heritage convention framework under the United Nations. There are two types of heritage: Cultural Heritage and the Natural Heritage (Kumar and Singh, 2015). UNESCO organizes information seminars and training workshops for capacity building, in countries with rich cultural and natural heritage, providing legal, operational, educational and awareness raising programs to protect the cultural property from illegal damage and trafficking (UNESCO, 2019a).

The first joint global strategy for capacity building for cultural heritage and natural heritage was held in Helsinki Finland. The outcome was to provide training to professionals to implement World Heritage convention in order to "manage world heritage properties and to strengthen technical, scientific, and traditional skills for the conservation of cultural and natural heritage" (ICCROM, 2019). It also focused on developing regional training strategies for capacity building of world heritage by developing (1) world heritage resources manual, (2) training and education on preparing nomination of heritage sites for world heritage title and periodic reporting, and (3) technical and management training on cultural heritage and natural heritage.

Spiritual Heritage Tourism Preservation, Protection, Promotion, and Management: SHT sites can be used for building capacity to preserve, conserve, and promote for economic prosperity and peace within and between the two nations. Research and education in heritage preservation, history, respect for all SHT sites and religious sites is required to keep them operational as living museums for students, scholars, and tourists, so that the SHT sites, places of worships, its surrounding gardens, spiritual souvenirs shops, and spiritual cultural ritual and festivals are well conserved and protected for the current and future generations. Research can also help to acquire the funding from government as well as private sector businesses and from individuals and non-government organizations volunteering help to provide expertise in protecting and preserving these SHT sites for peace building as well as for employment creation and income generation and becoming self-sustaining for preserving SHT thorough PPP programs. The main goal of any spiritual heritage site should be to respect it irrespective of any spiritual orientation and appreciate its historical, cultural, social, national, and local pride, educational, recreational, and economic value for current and future generations.

In terms of economic benefit and capacity building from SHT, it is essential for reaping the economic benefits from conservation, protection, and promotion of spiritual heritage sites, which is possible by preparing submissions for UNESCO world heritage listing. Garrod and Fyall's (2000) empirically recognized research on heritage tourism management concluded that management of heritage for tourism purposes depended upon eight elements: conservation, accessibility, education, relevance, local community, financial, recreation, and quality. The last three elements of financial, recreation, and quality are incorporated in the other five elements as seen later in Table 2.1. Therefore, capacity building from SHT faces many challenges particularly for India and Pakistan as they have thousands of years old numerous spiritual heritage sites belonging to eight religions. The two countries need to focus on cultural and natural spiritual heritage (1) conservation, (2) accessibility, (3) heritage education, (4) relevance, (5) local community, (6) financial and funding needs, (7) quality of heritage conservation and restoration work, and (8) spiritual heritage social, cultural, recreational, and economic values.

2.3 CASE OF SPIRITUAL HERITAGE TOURISM: INDIA AND PAKISTAN

The largest number of spiritual tourists are the domestic local population in India and Pakistan, the South Asian diaspora and also international tourists (Haq and Jackson, 2009; Kumar and Singh, 2015; Phukan et al., 2012; Qureshi, 2003; Timothy, 2011). Capacity building for conservation and preservation of cultural and natural SHT is an emerging profession and more and more people are being educated and made aware of preserving cultural and natural heritage monuments and sites for the future generation.

India: School of Planning and Architecture (SPA) in New Delhi, India is involved with ICCROM in training students and faculty about cultural heritage education, planning, implementing heritage preservation projects, natural habitat studies, and cultural heritage conservation and management in India. Natural disasters such as cyclones, floods, and earthquakes are the threats to cultural and spiritual heritage sites along with illegal land encroachment, excavations, and vandalism of the heritage properties by the people and visitors. India has numerous centuries old SHT sites and monuments from different religions. It has a total of 37 heritage sites out of which 29 are cultural, 7 natural and 1 mixed. Many of the Mughal Empire related spiritual, cultural, and natural heritage tourism sites, gardens and monuments are protected under UNESCO heritage such as the Agra Fort, Jaipur Fort, Humayun's Tomb, Red Fort, and the Taj Mahal. Numerous heritage monuments are under the Archaeological Society of India, which are not looked after well due to paucity of funds and are in danger. Further, many heritage tourism monuments, sites, gardens, places of worship are vandalized and dilapidated due to bad weather conditions. Furthermore, successive Indian governments in the last decade have the habit of changing names of historical heritage places and monuments, which is like destroying historical facts for the future generations.

According to UNESCO, the following natural and cultural heritage sites are listed for India. Natural Heritage: In this century, two sites are included under "Natural Heritage" namely Western Ghats in (2012) and the Great Himalayan National Park Conservation Area in (2014) and one in "Mixed Heritage" category called Khangchendzonga National Park in 2016.

Cultural Heritage: Nearly 13 monuments and sites are included from a total of 29 sites. They are (1) Mahabodhi temple complex in Bodh Gaya in 2002, (2) The Rock shelters of Bhimbetka in 2003, (3) Champaran Archaeological Park in 2004, (4) Victoria Railway Terminus in 2004, (5) The Red Fort complex in 2007, (6) Mountain Railways of India in 2008, (7) Jantar Mantar Jaipur in 2010, (8) Hill forts of Rajasthan in 2013, (9) The Queen's Step Well at Patan Gujarat in 2014, (10) Archaeological site of Mahavir-Jain Nalanda in Bihar in 2016, (11) Historic City of Ahmedabad in 2017, (12) The Architectural work of Le Corbusier in 2016, and (13) Victorian Gothic and Art Deco Ensembles of Mumbai in 2018.

Heritage Walled City of Ahmedabad: The sixth century historical walled city of Ahmedabad founded by Sultan Ahmed Shah in 1411 AD is listed and protected under the UNESCO world heritage in 2017 with its multicultural and multireligious characteristics of the timber architecture and famous for its town planning, sanitation, water storage, centre court yards, and living spaces (UNESCO, 2019b). Ahmedabad meets two (2 and 4) out of ten selection criteria to be listed as a world heritage site.

"The typology of the city's domestic architecture is presented and interpreted as an important example of regional architecture with a community-specific function and a family lifestyle that forms an important part of its heritage. The presence of institutions belonging to many religions (Hinduism, Islam, Buddhism, Jainism, Christianity, Zoroastrianism, Judaism) makes the historic urban structure of Ahmedabad an exceptional and unique example of multicultural coexistence" (UNESCO, 2019b).

According to UNESCO (2019b), heritage city of Ahmedabad an Islamic town established by the Sultan Ahmed Shah was classified on the basis of maintaining integrity, authenticity, and needs protection and management of heritage sites. In case of the integrity, heritage city of Ahmedabad has maintained its topography and geomorphology. Over the centuries, the hydrology, natural features, historic and contemporary built environment has been subjected to the changes and growth due to upgrade of infrastructure in terms of the city's population and community aspirations. Its infrastructure above and below ground has also been successively added and/or expanded with population growth. However, Ahmedabad city's gardens, open spaces, land use patterns, and spatial organization have largely remained unchanged for centuries. The authenticity of the unique wooden architecture and brick homes to suit its climate with the fort walls and three gates, Jama Masjid Mosque and large open spaces (Maidan), around the periphery of the mosque, small suburbs, and markets were established over time.

Due to the integrity and authenticity of the centuries old Ahmedabad Islamic city there was a need to protect and manage its cultural and natural heritage site. Therefore, 28 monuments listed by the Archaeological Survey of India along with 2696 heritage buildings and monuments identified by Ahmedabad Municipal Corporation were given heritage status for heritage conservation and sustainable management of its cultural and natural heritage, along with the 21st century urban development in and around the historic heritage city, and conserving and promoting its rich cultural heritage.

Pakistan: In Pakistan, similar to India, there are centuries old numerous cultural and natural heritage sites. These sites need to be preserved due to damage caused by plant/bio-growth, natural disasters together with illegal land encroachment, excavations, and vandalism of the heritage properties by the locals and visitors to the heritage sites. In Pakistan, UNESCO initiated the project called "Protection of World Heritage Sites of Pakistan through Capacity Building and Awareness Raising" to preserve the centuries old Buddhist caves and monasteries from Taxila and Islamic SHT sites from the ancient and medieval periods dating back to first to fifth century (UNESCO, 2019c).

In order to address the challenge of protecting spiritual tourism heritage, capacity building by PPP projects is the key to bring awareness by involving all stakeholders and the local community, UNESCO world heritage, government, NGO, industrialists, philanthropist, spiritual leaders, archaeology department, educational institutions, universities, architectural firms, academics from various disciplines such as scientist, historians, economist, and sustainable tourism management gurus and artists. In case of preserving the Buddhist world heritage site at Taxila, PPPs were formed between UNESCO, Islamabad government, Comsats Institute of Information Technology and, Hazara University to build capacity to solve the preservation challenges (UNESCO, 2019b).

Similarly, there are numerous spiritual tourism heritage sites in Pakistan related to other religions. Such as numerous centuries' old Hindu temples as Katsraj and over 550 years old Sikh Gurudhwara in Kartarpur Sahib, Sufi shrines, spiritual gardens, Zoroastrian fire temples, besides colonial churches in the various capital cities and hill stations. All these are in need of conservation and preservation for socioeconomic benefit of the future generations through capacity building for preservation and building peace in the minds of all people and protection from the natural disasters and illegal encroachment and vandalism of SHT.

In Pakistan, there is a treasure trove of heritage sites which are over 5000 years old. From the natural heritage of the Himalayan mountain ranges to the 5000 plus years old archaeological site of most ancient Indus valley civilization, civilizations along the Indus and Ravi and Chenab rivers, to centuries old Buddhist, Hindu, Sikh, and Islamic places of worship. In 2004, nearly 10 archaeological sites were listed tentatively and nomination as cultural heritage sites. There are nearly 26 cultural heritage and natural heritage sites considered for nomination to UNESCO world heritage listing such as forts, parks, monuments, mosques, and archaeological sites on the tentative listing. Of which there are six cultural heritage sites and monuments under UNESCO world heritage.

Cultural Heritage: Six sites are listed by UNESCO under the cultural heritage. (1) Archaeological ruins of Mohenjo-Daro in 1980, (2) the Buddhist ruins of Takht-i-Bahi and Sahr-i-Bahlol in 1980, (3) Taxila in 1980, (4) Lahore fort and Shalimar gardens in 1981, (5) Historical mountains of Makli Thatta in 1981, and (6) Rohtas fort in 1997.

Nominated Natural Heritage Sites: In 2016, eight national parks and landscapes were listed for nomination as UNESCO world heritage: (1) Derawar and the Desert Forts of Cholistan, (2) Hingol Cultural Landscape, (3) Karez System Cultural Landscape, (4) Nagarparkar Cultural Landscape, (5) Central Karakoram National Park, (6) Deosai National Park, (7) Ziarat Juniper Forest, and (8) The Salt Range and Khewra Salt Mine (UNESCO, 2019d).

Tentative List of Heritage Sites: There are 26 cultural and natural sites in Pakistan identified as a tentative list to be nominated and protected under UNESCO as world heritage. Many are SHT sites. For example, in Lahore: Badshahi Mosque, Wazir Khan's mosque, Tombs of Jahangir, son of the great Mughal emperor—Akbar, Hiran Minar in Sheikhupura, tomb of Sharukh-e-Alam tomb, and mosque Bibi Jawandi. Forts of Rani

Kot, Baltit, Derawar, and Desert Forts of Cholistan, Shah Jahan Mosque in Thatta, and Chaukhandi tomb in Karachi. Archaeological sites include, Mehrgarh, Rehman Dheri, Harappa, and Ranigat. Natural cultural heritage landscapes include Mansehra Rock, Hingol, Karez, and Nagarparkar; National Parks of central Karakorum, Deosai, Ziarat Juniper, salt ranges, and Khewra salt mines.

2.4 OPPORTUNITIES AND CHALLENGES FROM SHT CAPACITY BUILDING

Capacity building for SHT preservation, development, and promotion faces many opportunities and challenges as discussed below. Capacity building provides the key to build awareness regarding the socioeconomic, cultural, and environmental importance of protecting, conserving, and promoting spiritual heritage for the future generations. Capacity building brings together people with technical skills, historians, artists, and philanthropic financial support required for protecting and conserving centuries old multi-faith spiritual heritage in India and Pakistan, who share a common history for the benefit of the local population.

2.4.1 OPPORTUNITIES

i) Capacity building for SHT with involvement of all stakeholders' at local, national, and international levels for sustainable heritage tourism conservation and development for the future generations through world heritage education program, planning, implementation, and outcomes.
ii) Employment creation in various fields related to heritage conservation, preservation, tourism, and sustainable development of cultural and natural SHT sites for the benefit of the future generations.
iii) Planning, developing, implementing, and organizing SHT tours. Attracting local and overseas heritage tourists generates foreign exchange revenue and in turn having a multiplier effect on the economy for employment creation in various sectors, small and medium businesses supporting SHT-related goods and services.
iv) Opportunities for local, national, and international World Heritage Volunteers (WHV) activities on heritage sites for heritage tourism awareness of heritage sites, its importance to the community and economy, for conservation, preservation by students, youth

organizations, scholars, tourists, and people volunteering their skills, expertise, and knowledge in identifying, listing, educating, and building awareness, regarding heritage preservation, conservation, promotion, and preparing proposals for nominating cultural and natural sites for world spiritual heritage listing.

v) World heritage campaign activities and camps to empower local communities to be aware of cultural and natural heritage in their surrounding geographical areas. To empower to protect, conserve, preserve, and promote SHT sites. For example, in India WHV camps were held in Ahmedabad by Exilier foundation for Champaran-Pavagadh, Rani-Ki-Vav, or Queen's Stepwell and intercultural learning camp in Hampi, Karnataka.

vi) To raise funds for preserving SHT cultural or natural sites various fund-raising activities can be organized such as intercultural events, marathon run, sporting events, music, cultural, and film festivals at the heritage sites.

vii) Given the social, cultural, recreational, and economic significance of SHT sites many visitors also experience a spiritual transformation in their outlook for cultural and natural spiritual heritage and artifacts belonging to eight major spiritual faiths in India and Pakistan. Further, South Asian diaspora, students, researchers, and scholars' main purpose is to have transformational heritage experience, volunteer in spiritual heritage education, fund-raising for conservation, preservation, and promotion of spiritual heritage.

2.4.2 CHALLENGES

There are many challenges socioeconomic, cultural, political, religious, regulatory, population pressure, pollution, traffic, infrastructure development, and environmental problems faced by the two countries in capacity building for effective protection, conservation, and management of its natural and SHT. Capacity building for heritage tourism awareness plan has to be implemented as many spiritual heritage tourism buildings, sites, open spaces, and gardens are within the fort walls and city complexes of heritage buildings.

i) Heritage management planning with all the stakeholders is a challenge who have different goals and objectives. It is an essential instrument for conservation, preservation, and sustainable management of the SHT places of worship, spiritual gardens, and natural spiritual heritage (mountains, trees, statues, oceans, and caves).

ii) Listing and documenting of all categories of historical, cultural, natural, and SHT sites should be made by the local authorities and archaeological societies with the assistance from all stakeholders such as history, archaeology, and tourism students, educational institutions, historian, scholars of architecture, history, botanist, artists, builders, tour guides, and tourists for conservation, preservation and management of spiritual heritage buildings, monuments, sites, open spaces, and gardens.

iii) Implementation of a heritage management plan should be done effectively. For example, the development and construction of new infrastructure, buildings, roads, and bridges should not displace, demolish, and destroy cultural and natural heritage buildings, places of worship, and heritage open spaces and landscapes, and maintain their integrity and authenticity.

iv) Capacity building for heritage education and awareness for the locals and tourists/visitors should be prepared with involvement of all stakeholders and community for approval, implemented for sustainable SHT development.

v) Challenge of implementing heritage management plan due to corruption as funds allocated for preservation of heritage are not used effectively. Heritage protection, conservation, preservation, and development should take place within the heritage management plan criteria for conservation of heritage sites.

vi) The challenge is to incorporate the roles, regulation, and responsibilities of all visitors to protect the heritage sites and appreciate them with respect and not vandalize or damage the SHT sites.

vii) Challenge of capacity building by sharing all technical knowledge, skills, advice, and financial assistance from all over the world is required to preserve/restore SHT sites, buildings, places of worships, gardens, and nature to benefit the future generations, which are in danger of being destroyed.

viii) Challenge of war and terror which can destroy centuries old heritage forts, monuments, and spiritual heritage sites. For example, centuries old Buddhist SHT sites were destroyed due to war in many countries such as in Afghanistan or letting it decay or destroyed as it belongs to other minority religious groups.

ix) The conflict between India and Pakistan makes it also difficult to use scarce resources for capacity building and protect thousands of centuries old heritage cultural, archaeological, and natural sites. Both countries share centuries old Indus Valley civilization, cultural and natural heritage along with rich multi-spiritual and cultural heritage of places of worship. For example, various forts, and temples and mosques built within the fort, which are overgrown with vegetation and thus dilapidating, gardens and other heritage architecture which forms part of the spiritus heritage tourism. Such as the nominated cultural/SHT sites of Badshahi mosque in Lahore, various Sufi saints tombs in Multan, Buddhist and Hindu temples. Similarly, there are numerous centuries old forts, mosques, temples, and places of workshops belonging to eight religions in India which need capacity building with all stakeholders for conservation and restoration for future generations for sustainable SHT development.

x) Global Public–Private Partnerships for SHT (G-PPPs) conservation and development is required by working together with UNESCO, archaeological heritage societies for capacity building, fund raising, volunteering skills, and protecting and restoring world heritage sites like the centuries old numerous places of worships, forts, and other heritage and archaeological sites, centuries old monuments, wildlife, mountains, rivers, and forests in India and Pakistan.

2.5 PUBLIC–PRIVATE PARTNERSHIPS

PPPs framework between the key stakeholders within and between the countries for capacity building in world SHT monuments—cultural and natural sites is required. Under the United Nations Educational, Scientific and Cultural Organization UNESCO framework, partnerships are considered as the key driver or enabler for solving global problems and challenges. Innovative partnerships such as PPPs national and global between the key stakeholders such as building capacity for SHT conservation and preservation for the future generation is required to help to meet global, national, regional, and local challenges to bring about a sustainable development change and a long-term impact for the community (Haq and Medhekar, 2018, 2017). Through partnerships expertise, knowledge, resources, skills, and competencies required for capacity building can be pulled together for building capacity for SHT conservation and preservation having a sustainable long-term impact of the implemented projects and programs of capacity building for SHT conservation and preservation for socioeconomic benefit of the future generations.

The chapter proposes that the PPPs for SHT committee in case of India and Pakistan should consist of UNESCO representatives from world heritage Paris, cultural and management of heritage UNESCO sites, SHT representatives from the two countries, skilled architects in conservation and preservation, traditional artists, besides educational institutions, faculties of architecture, business and SHT management, local government, local NGO's, and religious owners and protectors of SHT sites.

The first step is to start from schools, universities, and educational institutions. Partnerships between spiritual heritage archaeological sites, state and educational institutions will help to preserve SHT sites (Tilson, 2005). Following steps can be taken in case of India and Pakistan for capacity building in conserving and preserving SHT in both the countries, given the shared history and culture and nearly eight major religions namely: Buddhism, Christianity, Hinduism, Islam, Jainism, Judaism, Sikhism, and Zoroastrian represented in these two countries with their centuries old religious/SHT cultural and natural sites.

i) Training young minds at school and universities regarding preservation and conservation of world heritage which includes natural and cultural, and SHT being part of cultural heritage in the curriculum, such as national and local "Heritage in Box" toolkit project, similar to the one organized in Mongolia by UNESCO.
ii) Capacity building workshops "Heritage in Box" toolkit project can be extended by involving school students and university alumni as promoters for building awareness among the society for preserving SHT sites for the future generation and for reaping social-economic benefit from such SHT capacity building and preservation for sustainable socioeconomic impact. This requires capacity building and respect for all SHT sites—cultural monuments and natural parks and sites, with the aim to protect by having interreligious/intercultural dialogue for long-term sustainable development and peace.
iii) Capacity building via education and skill development, training for SHT ethics, and respect for all heritage cultural and natural parks and sites are required for preparing successful nomination of SHT sites for world heritage listing by UNESCO.

iv) Capacity building through education and training in periodic reporting to preserve SHT sites for economic prosperity, promotion, and peace between India and Pakistan. Capacity building through educational workshops and training to preserve the original names, images, historical brand of heritage places, and monuments.

v) Capacity building at local, national, and global level for policing and being a neighborhood watch dog for protecting the SHT cultural and natural sites and monuments from being destroyed by nature or vandalized by illegal land encroachment and people and visitors.

vi) In recent years in India, names of heritage monuments of cultural and natural significance is changed by the government. For example, the name of Victoria railway terminus in Bombay was changed to Chhatrapati Shivaji terminus. Similarly, the government in power wants to change the name of Aligarh city to Prayag, its Aligarh Muslim University which was founded in 1875 by Sir Syed Ahmed Khan and Aligarh railway station etc. to Prayag or Harigrah. A forest area was cleared, and a city called Aligarh was founded and established by the Great Mughal emperor Akbar. The future generations to come will not know the true history of the monument. It is essential that academic scholars and historians should write and portray the true history of the country irrespective of the colonial past or the religious affiliations of the rulers who built and named places and cities and also built forts and monuments. The government should preserve the significance of historical names given for future generations, and not change the names of historical places, heritage parks and monuments.

Haq and Medhekar (2018) have identified challenges in capacity-building partnership and management of heritage tourism for India and Pakistan in context of: conservation, accessibility, relevance, finance, quality of heritage conservation work, heritage education for awareness, recreational use of heritage sites for music and cultural festivals, and local community driving SHT protection, promotion, conservation, and preparing nominations with global PPPs for UNESCO listing as world spiritus heritage. These are important for capacity building from SHT preservation, protection, and promotion involving the various stakeholders. A recent example of peace and capacity building partnership through spiritual heritage tourism at the government to government level between India and

Pakistan is to bring about Sikh heritage awareness, protection, preservation, development, and promotion with the opening of 'Kartarpur Spiritual Heritage Tourism Corridor' on the 9th of November 2019, to celebrate 550 birth anniversary of the founder of the Sikh religion. Kartarpur Sikh Gurdwara (temple) is a very sacred heritage place, site and spiritual monument to the Punjabi Sikhs and Punjabi Muslims alike. Kartarpur heritage site and temple is like pilgrimage to Mecca to the Sikhs from all over the world, being the resting place of the founder of Sikhism—Guru Nanak, where he spent the last 20 years of his life (Akhtar et al., 2019). Kartarpur spiritual heritage tourism corridor is focusing on the foundation for peace and capacity building initiative for conservation of centuries old spiritual heritage, besides the economic multiplier effect of socio-economic opportunities for bilateral travel, tourism and trade for the benefit of the two countries. Table 2.1 provides the challenges in capacity-building global partnership for planning, management, implementation, and promotion of SHT innovation for capacity building in India and Pakistan.

TABLE 2.1 Capacity-Building Partnership Challenges from Spiritual Heritage Tourism.

Heritage tourism challenges	Methodological approach for spiritual heritage tourism innovation for capacity building
Conservation	• Driving forces—heritage awareness, protection, preservation, development, and promotion. • Heritage education, expertise, and innovative skills. • Availability of finance. • Technological innovations. • Quality of heritage conservation work. • Economic performance and progress in identifying, preserving, protecting, promoting, and preparing submissions for spiritual heritage listing by UNESCO as cultural or natural world heritage.
Accessibility	• Driving forces—heritage awareness, protection, preservation, development and promotion. • Heritage education, expertise, and innovative technical skills. • Availability of finance. • Quality of heritage conservation work. • Technological innovations.

TABLE 2.1 *(Continued)*

Heritage tourism challenges	Methodological approach for spiritual heritage tourism innovation for capacity building
Relevance	• Driving forces—heritage awareness, protection, preservation, development and promotion. • Heritage education, expertise, and innovative skills. • Quality of heritage conservation work. • Technological innovations. • Cultural, social, recreational, and economic contribution.
Education	• Driving forces—heritage awareness, protection, preservation development and promotion. • Heritage education, expertise, and innovative skills. • Availability of finance. • Technological innovations. • Quality of heritage conservation work. • Economic performance and progress in identifying, preserving, protecting, promoting, and preparing submissions for spiritual heritage listing by UNESCO as cultural or natural world heritage. • Innovative transformation for capacity building and spiritual heritage conservation.
Local Community	• Driving forces—heritage awareness, protection, preservation, development and promotion. • Heritage education, expertise, and innovative skills. • Technological innovations. • Quality of heritage conservation work. • Cultural, social, recreational, and economic contribution. • Local and global PPPs for capacity building • Economic performance and progress in identifying, preserving, protecting, promoting, and preparing submissions for spiritual heritage listing by UNESCO as cultural or natural world heritage.

Source: Developed for this chapter and adapted from Haq and Medhekar (2018).

2.6 CONCLUSIONS

This chapter has highlighted the need for capacity building between India and Pakistan in partnerships in order to conserve, preserve, and promote

SHT sites and spiritual cultural heritage which is thousands of years old for economic prosperity, progress, and peace between the two countries at all levels people to people, government to government, and at the diplomatic level to bring about sustainable transformation between the two countries. The intergenerational socioeconomic benefits in capacity building for SHT development, conservation, and preservation are far greater for local communities in terms of job creation, growth in national income, foreign exchange earned, than the costs of neglect and destroying centuries old SHT sites based on religious divide and communal riots. It is the government's archaeological societies' responsibility to list the SHT sites belonging to all religions, no matter how small or significant, in the two countries and preserve, protect, and promote them for economic prosperity, communal harmony, all inclusivism, and peace within and between the two countries.

Government, educational institutions, private sector businesses, and the local community along with global partners like UNESCO, developed countries, other private charitable and philanthropic individuals and organizations are the key driving force for capacity building via educational institutions, and participatory workshops and training necessary for any preservation, protection, and promotion to bring about innovative transformation of the centuries old SHT sites of all religions for socioeconomic benefit, economic prosperity, and peace between India and Pakistan.

Capacity building is possible by implementing heritage protection policy without bureaucratic delay and corruption, building infrastructure for access, appointing local educated people, and expertise with knowledge of built heritage and history of the local areas and development of heritage conservation skills through educational institutions. However, G-PPPs are essential for capacity building on one hand to conserve and protect heritage sites with "*innovative capacity-building transformation strategies*" for empowering local communities to conserve, preserving the SHT for the future generations. On the other hand, to ensure that "SHT *development*" is based on identifying and mapping "*spiritual heritage tourism circuits*" (SHTC), places of worships, and monuments from a localized knowledge perspective for protection, conservation, restoration, transformation, and promotion, in a responsible manner so that SHT sites are not neglected as "belong to the other" or vandalized by the locals in communal riots or by the tourists.

Future research direction can apply Lempert (2015) classification of historical sites for India and Pakistan. Local educational institutions with community and local council's involvement could take up the public consultation and grassroots participation of the local stakeholders meeting, project to identify and classify heritage sites based on generic heritage, abandoned, deteriorated, and restored/conserved and the transformation stage for reuse as *heritage museums* along with *responsible heritage tourism development*.

Secondly, framework from Table 2.1 can be empirically tested qualitative and quantitatively to determine the factors that drive spiritual heritage conservation, preservation, and promotion in India and Pakistan. The need of the 21st century in both the countries is to conserve and protect the spiritual heritage sites by adopting responsible innovative transformational strategies, respect for all religious/spiritual sites amidst rapid demolition of the heritage sites for land clearing, for commercial development by builders. Local communities backed by educational institutions, funding bodies, UNESCO, local businesses, and government should take pride in preserving their heritage and culture and transform their lives for the benefit of the future generations. As they will be directly committed to preserving their tangible and intangible heritage, to reap the social, cultural, recreational, and economic benefits from spiritual heritage conservation, fund raising, holding music, interfaith and cultural festival, experience innovative social transformation, leading to sustainable development of centuries old SHT circuits and clusters in India and Pakistan for the future generations. This will also bring about an innovative transformation in the visitors (tourists, students, volunteers, funding bodies, and scholars) to build capacity and push for heritage conservation and rejuvenation by preparing submissions for world heritage UNESCO listing of the SHT sites in India and Pakistan.

KEYWORDS

- **spiritual heritage tourism**
- **capacity building**
- **preservation**
- **protection and peace**
- **regional development**
- **public–private partnerships**

REFERENCES

Akhtar, M. S.; Jathol, I.; Hussain, Q. A. Peace Building through Religious Tourism in Pakistan: A Case Study of Kartarpur Corridor. *Pakistan Social Sci. Rev.* **2019,** 3(2), 204–212.

Bandyopadhyay, R.; Morais, D. D.; Chick, G. Religion and Identity in India's Heritage Tourism. *Ann. Tour. Res.* **2008,** *35*(3), 790–808.

Chakraborty, S. Developing an Integrated Tourism Circuit in Gar Mandaran-Kamarpukur-Joyrambati in the State of West Bengal, India. *South Asian J. Tour. Heritage* **2011,** 4(1), 144–155.

Garrod, B.; Fyall, A. Managing Heritage Tourism. *Ann. Tour. Res.* **2000,** *27*(3), 682–708.

Haq, F.; Jackson, J. Spiritual Journey to Hajj: Australian and Pakistani Experience and Expectations. *J. Manage. Spiritual. Relig.* **2009,** *6*(2), 141–156.

Haq, F.; Medhekar, A. Branding Spiritual Tourism as an Innovation for Peace between India and Pakistan. *IJSEI* **2013,** *2*(5), 404–414.

Haq, F., Medhekar, A. Spiritual Tourism between India and Pakistan: A Framework for Business Opportunities and Threats. *World J. Soc. Sci.* **2015,** *5*(2), 190–200.

Haq, F.; Medhekar, A. Is Spiritual Tourism an Innovation in Tourism for India and Pakistan. In *Business and Economics: Leadership, Innovation and Entrepreneurship as Driving Forces of the Global Economy*; Benlamri, R., Sparer, M., Eds.; Springer International: Switzerland, 2017; Chapter 44, pp 519–530.

Haq, F.; Medhekar, A. Challenges for "Innovative Transformation" in Heritage Tourism Development in India and Pakistan. In *Conservation and Promotion of Heritage Tourism*; Srivastava, S., Ed.; IGI Global Publishing: Penn, USA, 2018; pp 127–154. https://www.igi-global.com/gateway/chapter/213400.

ICCROM. World Heritage Capacity Building, 2019. https://www.iccrom.org/section/world-heritage/world-heritage-capacity-building.

Kumar, A.; Singh, P. Cultural Heritage Tourism of Ayodhya-Faizabad. *Geographer* **2015,** *62*(2), 66–74.

Lempert, D. Classifying Heritage by States of Decay, Restoration, and Transformation for Tourism, Teaching, and Research: "un-freezing" Sites in Time to Reveal Additional Meanings. *J. Heritage Tour.* **2015,** *10*(4), 378–398.

Medhekar, A.; Haq, F. Growth of Spiritual Tourism in India: A Public Private Partnership Approach. *Int. J. Busi. Intell. Manage.***2009,** *1*, 1–12.

Phukan, H.; Rahman, Z.; Devdutt, P. Emergence of Spiritual Tourism in India. *Int. J. Market. Technol.* **2012,** *2*(4), 300–317.

Qureshi, R. B. Lineage, Shrine, Qawwali, and Study Circle: Spiritual Kinship in Transnational Sufism. *Relig. Stud. Theol.* **2003,** *22*(1), 63–84.

Seraphim, J.; Haq, F. Challenges for Promotion of Heritage Tourism: Case Study of the UAE. In *Conservation and Promotion of Heritage Tourism*; Srivastava, S., Ed.; IGI Global Publishing: Penn, USA, 2018; pp 35–56. https://www.igi-global.com/gateway/chapter/213400.

Sharpley, R.; Sundaram, P. Tourism: A Sacred Journey? The Case of Ashram Tourism, India. *Int. J. Tour. Res.* **2005,** *7*(1), 161–171.

Shinde, K. A. Visiting Scared Sites in India: Religious Tourism or Pilgrimage? In *Religious Tourism and Pilgrimage Management: An International Perspective*; Raj, R., Morpeth, N. D., Eds.; CAB International: UK, 2007; Chapter 15, pp 170–183.

Simone-Charters, M. T.; Boyd, S. W. Development of Religious Heritage Tourism in Northern Ireland: Opportunities, Benefits and Obstacles. *Tourism* **2010,** *58*(3), 229–257.

Tilson, D. J. Religious-Spiritual Tourism and Promotional Campaigning: A Church-State Partnership for St. James and Spain. *J. Hosp. Leisure Market* **2005,** *12*(1–2), 9–40.

Timothy, D. J. *Cultural Heritage and Tourism: An Introduction*. Channel View Publications: Bristol, 2011.

Timothy, D. J.; Boyd, S. W. *Heritage Tourism*; Prentice Hall: Harlow, 2003.

Timothy, D. J.; Olsen, D. H. Conclusion: Whither Religious Tourism? In *Tourism, Religion & Spiritual Journeys*; Timothy, D. J., Olsen, D. H.; Eds.; Routledge: Oxford, UK, 2006; pp 271–278.

UNESCO, 2019a. http://www.unesco.org/new/en/culture/themes/illicit-trafficking-of-cultural-property/capacity-building/.

UNESCO. Historic City of Ahmedabad—World Heritage Centre UNESCO, 2019b. http://whc.unesco.org/en/list/1551.

UNESCO. Protecting World Heritage of Pakistan through Awareness Raising and Capacity Building, 2019c. http://unesco.org.pk/culture/capacity_building.html.

UNESCO. Pakistan UNESO World Heritage Listing, 2019d. http://whc.unesco.org/en/statesparties/pk.

CHAPTER 3

STRATEGIES FOR RURAL TOURISM AND THE RECOVERY OF CULTURAL HERITAGE: CASE STUDY OF LA VERA, SPAIN

FRANCISCO JAVIER CASTELLANO-ÁLVAREZ[1],
MARÍA DE LA CRUZ DEL RIO-RAMA[2],
AMADOR DURÁN-SÁNCHEZ[3], and JOSÉ ALVAREZ-GARCÍA[3*]

[1]*Department of Economy, University of Extremadura, Cáceres, Spain*

[2]*Department of Business Management and Marketing, University of Vigo, Vigo, Spain*

[3]*Department of Financial Economy and Accounting, University of Extremadura, Cáceres, Spain*

[*]*Corresponding author. E-mail: pepealvarez@unex.es*

ABSTRACT

In the mid-eighties, the European institutions show their growing concern for the economic and social cohesion of the territory and the consequences that the imminent reforms of the Common Agricultural Policy (CAP) could have on the rural environment.

In this context, the LEADER Initiative (acronyms that correspond to the French term Liaison Entre Actions de Développement de l'Economie Rurale) and, in the Spanish case, the PRODER Program (Program of Development and Economic Diversification of Rural Areas) aspire to create employment in alternative sectors to agriculture with the objective of retaining, in rural areas, the population expelled from the primary sector.

Within these programs and based on a local area of action, the different Local Action Groups (LAGs) must define their rural development strategies. This is when many rural spaces turn the rural tourism sector into an axis of development. Often, tourism investments will be accompanied by other support actions aimed at the recovery of various elements of cultural heritage.

Using the methodology of the case study, and based on the example of the region of La Vera (Extremadura, Spain), this research aims to analyze the practical application of this type of program and, especially, the strategy and investments directed to the promotion of rural tourism. In this endeavor, special attention will be given to several issues such as (1) the distribution of the investments made and the implicit objectives to them; (2) the involvement of the local population in the promotion of projects associated with rural tourism; and (3) the feasibility, in the long term, of the tourism projects promoted.

Understanding, in its just terms, of the tourism development strategy of the region of La Vera will require us to analyze those other actions, which regardless of the resources allocated to the promotion of rural tourism are aimed at enhancing the tourist resources of the area with various lines of action among which the recovery of heritage.

The research provides interesting conclusions about the projects executed, the involvement of local promoters in the realization of tourism investments, and the orientation of public investments aimed at the recovery of cultural heritage.

3.1 INTRODUCTION

In 1986, the Single European Act (in Article 118) considers tourism as one of the sectors that could contribute the most to the development of the rural environment and the cohesion of the territory. It was understood, even then, that rural tourism could be a useful instrument to compensate for the decrease in agrarian income derived from the imminent reforms of the Common Agricultural Policy (CAP). These are reforms that the European Commission outlines through various statements, such as the reports on the "Perspectives on the CAP" (1985), "The future of the rural world" (1988), or the one known as "Agenda 2000" (1997). In these documents, it is recognized that although agriculture was still of great importance in rural areas, it could no longer articulate the rural economy (Pisani, 1994), and there is a commitment to the economic diversification of the rural environment.

In this context, rural development policies conceive rural tourism as "tourism with a difference", an alternative to traditional "sun and beach" tourism. The promotion of this activity linked to agricultural holdings would lead to occupy part of the labor surpluses of farms and could become an ideal supplement to farmers' income. In addition, the use and recovery of old buildings enable them to perform this activity without having to resort to the farmer's indebtedness to make large investments.

It is then when through several calls (1991, 1994, 2000) that the European Commission promotes the Leader Initiative, a novel proposal whose aim is to promote endogenous and sustainable development of rural areas. The promotion of rural tourism will thus become one of the main axes of development underlying these programs.

The response to the first calls of the Leader Initiative was so positive that many territories that were interested in its application could not have access to it. To respond to this demand, some countries approved similar programs inspired by the model proposed by the Leader. This is the case of the Spanish authorities, who through their Ministry of Agriculture approved two editions of the PRODER Program (1996; 2002).

Although, initially, the interest of Leader and PRODER in rural tourism was linked to agricultural operations, several investigations, such as those by Arnalte and Oliveira (2009) or Pulido and Cárdenas (2011), show that the presence of farmers among the promoters of tourism investments will be symbolic.

In any case, the application of rural development programs during the 1990s marks a milestone in the emergence of a sector, which at the beginning of that decade hardly had a testimonial presence in rural areas. What happened in Extremadura (which is the region where the case study of this research is located) is a good proof of this: from 1993 to 2010, more than 600 tourist establishments were created in rural Extremadura (Campesino, Sánchez, and Rengifo, 2010). According to the estimates of Mejías (2006) in this region, during the second half of the 1990s, due to the Leader II Initiative and the PRODER I Program, more than 3500 accommodation places were created, and around 80% of the beds offered in Extremadura. However, despite these achievements, we must be cautious when pondering the impact of these programs on the national aggregate of the tourism sector; in fact, in 2003, the accommodation places offered by rural areas barely accounted for 2% of the total of the Spanish tourism sector (Alario, 2004).

Under the definition of "rural tourism", depending on the motivation for demand, different ways of practicing tourism can be differentiated. Ivars (2000) highlights the following: agrotourism, interior tourism, ecotourism, cultural tourism, etc. Ruíz (1995) considers that a number of variables influence the expectations that the tourist activity generates in the rural world, such as (1) saturation of sun and beach tourist areas; (2) growing demand for tourism related to the environment, cultural heritage, or sports; (3) positive (even idyllic) image that the media conveys of the rural way of life; and (4) new trends consisting of short stays in nearby areas.

However, the fact that a certain territory allocates a large amount of resources to the promotion of rural tourism will not be enough to ensure its success. Not all rural areas have the same potential to promote a competitive tourism sector. Before the execution of this type of investment, the natural, cultural, economic, and human resources of the territory must be taken into account.

Although tourism can contribute to the development of the most backward areas and the cohesion of the community territory, Millán (2002) reflects the difficulties involved in promoting quality tourism in spaces characterized by demographic regression, their aging population, or their low-skilled workforce. In these areas, even the existence of great tourist attractions does not guarantee the success of the policies that promote this activity. For the aforementioned author, when a territory undertakes a tourism development strategy, it must

a) avoid tourist congestion in rural areas;
b) achieve to protect them within management policies that propose a comprehensive vision of the territory;
c) train the local workers to perform the different activities in this sector;
d) create a wide range of leisure alternatives; and
e) offer an adequate and attractive image of the space that is to be promoted.

Despite their limited resources and their clear subsidiary nature regarding the regional policy or agrarian policy, rural development programs achieved remarkable results and aroused the illusions of the majority of Europe's rural environment. The explanation for this apparent contradiction between resources, results, and expectations is found in the methodology applied. The peculiarities of this development model have been the object of numerous investigations, both from a theoretical and empirical point of view (Zapatero and Sánchez, 1999; Garrido and Moyano, 2002; Cebrián, 2003; Nieto and Gurría, 2010; Ramos and Garrido, 2014; Navarro, Woods and Cejudo, 2016; Nieto and Cárdenas, 2017).

An appropriate definition of the territory is essential when applying these programs. At local (too small) or regional (very broad) level, the region is understood as "that territorial area homogeneous enough to share problems and solutions" (Guiberteau, 2002, p. 95). Any analysis that aspires to deepen the knowledge of rural development programs and strategies to promote tourism should take a regional area of activity as reference.

Given some previous ideas regarding the relevance that the promotion of tourism acquires within rural development programs, the following section specifies the objectives of the research and details its methodological aspects. In the third section, bearing in mind the aforementioned objectives, the results of the investigation are addressed; and, finally, in the fourth section, its most relevant conclusions are shown.

3.2 OBJECTIVE AND METHODOLOGY OF THE RESEARCH

3.2.1 OBJECTIVE OF THE INVESTIGATION

Based on the methodology of the case study, this paper aims to analyze the practical application of rural development programs and, in particular, the strategy and investments aimed at promoting rural tourism. In this endeavor, the following will be studied: (1) projects promoted by private investors and (2) actions led by public entities.

The following will be studied regarding private promoters: (1) distribution and objectives of their investments and (2) participation of local stakeholders in the promotion of tourism projects.

Regarding public actions, complementary investments aimed at promoting rural tourism will be analyzed, as well as those that apart from the aforementioned measure, seek to recover rural heritage elements.

3.2.2 THE REGION OF LA VERA AS AN OBJECT OF THE CASE STUDY

Within a region such as Extremadura, which in itself constitutes an ideal scenario to analyze the application of rural development programs

(González, 2006), La Vera has a series of characteristics that make it be considered a paradigmatic study of the case. The choice of the case is a decision of great importance, which determines the success of this type of investigation. From the methodological point of view, it is essential:

a) For the chosen case to have its limits clear.
Located in the Northeast of the province of Cáceres, with an area of 885.98 km^2, and composed of a total of 19 municipalities, the region of La Vera is bordered to the North by the Southern tip of the Sierra de Gredos and the Valle del Jerte, to the East by the provinces of Ávila and Toledo, to the South by the Tiétar river; and to the West by the region of Plasencia (Fig. 3.1).

FIGURE 3.1 Limits and location of the scope of the study.

b) For it to be a relevant case, valid to contrast what one wishes to study.
In La Vera, there are great natural and heritage resources that boost rural tourism to become one of the main axes of development of the region.

Sierra de Gredos should be highlighted among the natural resources, which, depending on its altitude, has a great wealth of flora and fauna; its foothills characterize the physiography of La Vera. Some of its most emblematic peaks are La Covacha (Losar), Los Infiernillos (Guijo de Santa Bárbara), El Cancho (Talaveruela), and La Panera (Aldeanueva).

On the other hand, its hydrographic network, composed of numerous Gorges, with pure and crystalline waters, constitutes a great tourist attraction of the area. The embedding of these Gorges in the granitic bedrock of Gredos leads to the formation of natural "pools", waterfalls, and bathing areas, which become a leading tourist resource in the summer season. Some of these areas have their own particular names: Las Pilas (Collado), El Trabuquete and Estaca (Guijo de Santa Bárbara), El Lago (Jaraíz), Puente Jaranda (Jarandilla), Cuartos (Losar), El Tobogán or el Calambuco (Villanueva), etc.

The landscape wealth of the region of La Vera is also a consequence of its climatology. The influence of the Sierra de Gredos conditions the climate of La Vera, creating a Mediterranean microclimate characterized by mild temperatures and abundant humidity.

Remarkable cultural resources should be added to the tourist potentialities that the natural resources offer, among which the following stand out: (1) festivities declared of regional tourist interest such as Los Empalaos in Valverde or El Peropalo in Villanueva and (2) an enormous architectural and cultural heritage, distributed throughout the region, and whose best examples are the five municipalities declared Historic–Artistic Sites: Cuacos de Yuste, Garganta la Olla, Pasarón, Valverde, and Villanueva.

c) For it to be a representative case of the reality that one wishes to study:

The evolution of this region is representative of the trend followed by the whole rural area of Extremadura and Spain. Like many other rural areas, since the 1980s, La Vera has had to face a real structural change in circumstances where the economy of La Vera has been characterized by its low diversification, excessive dependence on employment and income from the primary sector, low-skilled human resources, considerable deficiencies in terms of infrastructure, unemployment rates above the regional and national average, etc.

3.2.3 APPROACH OF THE INVESTIGATION AND SELECTION OF THE SAMPLE

In relation to the time horizon of research, studying a rural development strategy requires perspective. Therefore, the proposed analysis will refer to a sufficiently long period so that these types of programs can be effective and far back in time to analyze the viability that the projects driven by it may have had in the long term.

Therefore, the temporal scope of the research will cover the two editions in which the PRODER Program was applied; that is, from the second half of the 1990s to the first half of the 2000s; a decade seems a reasonable period of time so that the action of these Programs can be materialized on the territory.

In their development strategies, these programs aim to take advantage of all the productive activities that are carried out, or can be carried out, on the territory. This multisectoral nature is transferred to the structure of the program. Although there are differences in the denomination of the different measures in the two editions of the PRODER Program, Table 3.1, which is based on their purpose, groups them together and reflects the investment committed in each of them:

TABLE 3.1 Distribution by Investment Measures made in PRODER I and II in La Vera.

	PRODER I	%	PRODER II	%
CEDER operation and technical assistance	608,532.98	11	666,967.20	12
Recovery and conservation of the environment	1,227,340.71	23	1,097,471.54	20
Total non-productive measures	**1,835,873.69**	**34**	**1,764,438.74**	**32**
Promotion of rural tourism	2,157,490.16	40	2,072,606.99	38
SMEs, crafts, and services	943,459.43	17	722,488.60	14
Valorization of agricultural production	483,272.34	9	886,140.89	16
Total productive measures	**3,584,221.93**	**66**	**3,681,236.48**	**68**

Source: Own elaboration based on the data provided by ADICOVER.

Due to the productive nature of the projects implemented within the measure of promotion of rural tourism, most of the investments made within this line of action will have a promoter and will be of private nature. In relation to these projects, to address some of the objectives of the research, it was necessary to resort to a qualitative methodology based on extensive field work, in which the main source of information was conducting interviews with the promoters of tourism projects.

When conducting interviews, to obtain a sample of the most interesting actions promoted by private developers, the following criteria were applied:

a) That the boosting and main source of financing of the action is private.
b) That the PRODER subsidy should have a minimum value of €12,000.
c) That PRODER's contribution should represent at least 20% of the final investment.

Table 3.2 classifies the private projects under analysis, details their investment, as well as the representativeness of the sample according to the criteria discussed.

TABLE 3.2 Classification of Private Projects and Representativeness of the Sample.

Typology of the projects	Total projects	Investment	Sample projects	Investment sample	%
Construction of rural accommodation	17	2,679,160.29	14	2,280,642.50	85.12
Modernization and equipment accommodation	10	658,302.90	5	604,596.11	91.84
Modernization restaurants	4	169,167.78	3	154,664.80	91.43
Complementary activities	3	65,896.81	1	32,721.91	49.65
Total	**34**	**3,572,527.78**	**23**	**3,072,595.32**	**86.01**

Source: Own elaboration.

Investigations based on case studies do not justify their representativeness based on conducting a large number of interviews. In this type of analysis, representativeness cannot be justified in statistical terms because the methodology itself limits the fieldwork to the case under investigation. The aforementioned representativeness must be justified in analytical terms, proving that "the case is appropriate for the type of theoretical discussion that is to be explained" (Coller, 2000, p. 56). Based on this premise, by conducting interviews with a sample of private promoters, it was possible to obtain the qualified opinions of those who directly participated in the management and execution of 86.01% of the investments committed in private projects. Finally, it should be noted that conducting the interviews at the place where the investments were made allowed for a better understanding of their purpose and an "*in situ*" evaluation of the projects included in the sample.

3.3 RESULTS OF THE INVESTIGATION

3.3.1 DISTRIBUTION OF INVESTMENTS CARRIED OUT

Table 3.3 shows the investments made within the tourism promotion strategy and shows the relative importance of each of their lines of action: (1) creation and modernization of rural accommodation; (2) restaurant businesses; (3) promotion of complementary activities for the tourist; and (4) transversal actions of a public nature.

In addition to private projects, within the rural tourism development strategy, it is worth highlighting the existence of a series of public actions, which are managed directly by the LAG, and are financed by the public resources of the program. Although the analysis of these actions is addressed in Section 3.3, their inclusion in the previous table allows for a global view of the lines of activities undertaken and their relative importance within the whole measure.

As shown in Table 3.3, the investment allocated to the creation and modernization of accommodation stands out over all other considerations. Perhaps, this can be an issue of debate since this type of action concentrates practically 80% of the total resources of the measure. Moreover, if these investments are added to those made in transversal actions of a public nature, only 5% of the resources are allocated to two essential issues for the development of the tourism sector in rural areas, which are restaurant businesses and promotion of complementary activities for tourists.

TABLE 3.3 Distribution of Investments in Rural Tourism Promotion by Lines of Action.

Typology of the projects	No. of projects	Investment	%
Creation and modernization of accommodation	27	3,337,463.19	78.90
Investments in restaurant businesses	4	169,167.78	4.00
Complementary activities	3	65,896.81	1.55
Transversal actions of a public nature	32	657,569.37	15.55
Total	66	4,230,097.11	

Source: Own elaboration.

3.3.2 INVOLVEMENT OF LOCAL STAKEHOLDERS IN THE PROMOTION OF TOURISM PROJECTS

The private projects referred to in this section will be those included in the sample, since knowing the origin of the promoters required asking this question in the personal interviews. To analyze this question, the following table differentiates the aforementioned promoters according to their origin.

To analyze this question, Table 3.4 differentiates private promoters according to their origin, based on this differentiation. Table 3.5 classifies the investments made according to the orientation of the projects. Due to the fact that regarding the origin of the promoters, there are significant differences between investments for the creation of accommodation and modernization, the aforementioned table differentiates these two types of projects.

TABLE 3.4 Classification of the Promoters of Tourism Projects According to their Origin.

Origin	Characteristics of the promoter
Natives	People who have and had their permanent residence in the region.
Returnees	People who, having their roots in the region, emigrated to urban environments. Now, they return to undertake tourism projects.
Neorrurals	People who, without having any previous relationship with the region, leave the urban environment to live in a rural area.

Source: Own elaboration.

TABLE 3.5 Origin and Investments Made by Private Promoters.

Typology of the projects	Failed	Natives	Returnees	Neorrurals
Accommodation				
Newly built accommodation	725,026.26	329,699.50	117,302.34	1,108,614.40
Accommodation modernization		604,596.11		
Restaurant business		154,664.80		
Complementary activities	32,721.91			
Total	**757,748.17**	**1,088,960.41**	**117,302.34**	1,108,614.40

Source: Own elaboration.

The analysis of the involvement of local promoters in tourism investments provides some ideas of particular interest. In the first place, reference must be made to the fact that failed investments amount to one-third of the total investment made by private promoters. In addition, as shown in the previous table, unsuccessful projects are concentrated in a specific type of action: newly built rural accommodation. It is also striking that the only project related to the performance of complementary activities is also unsuccessful.

The analysis of the involvement of local promoters in tourism investments provides some ideas of particular interest. In the first place, reference must be made to the fact that failed investments amount to one-third of the total investment made by private promoters. In addition, as shown in the previous table, unsuccessful projects are concentrated in a specific type of action: newly built rural accommodation. It is also striking that the only project related to the performance of complementary activities is also unsuccessful.

As Graph 3.1 shows, in the materialization of tourism investments, two types of promoters stand out: native and neorrural. However, the activities in which they materialize their investments are very different; while the promoters of neorrural origin aim all their projects at the creation of new rural accommodation. Native promoters hardly commit 30% of their investments to this type of activity.

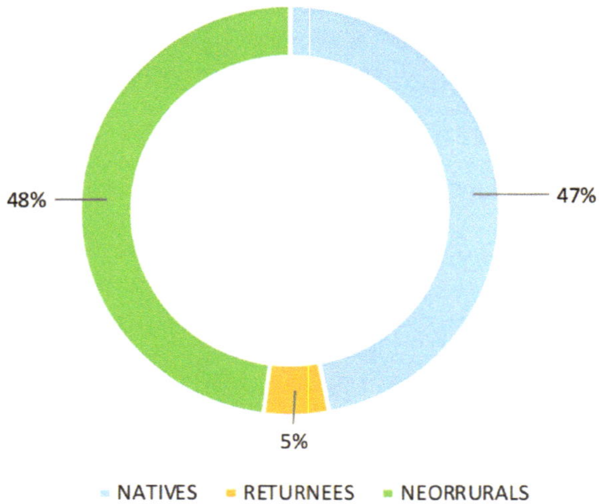

GRAPH 3.1 Investments made according to the origin of the developer.
Source: Own elaboration.

All investments aimed at the modernization of rural accommodation or restaurant businesses were addressed by the local promoters. Therefore, the investments of neorrurals and returnees are concentrated in the new creation of rural accommodation, which in addition, is the type of project in which there are more failed investments. The possible inter-relationships between these two facts could constitute of future lines of research.

The following graph complements the data shown in Table 3.5 and in Graph 3.1 and reflects the significant concentration of tourism investments in a single type of activity (Graph 3.2).

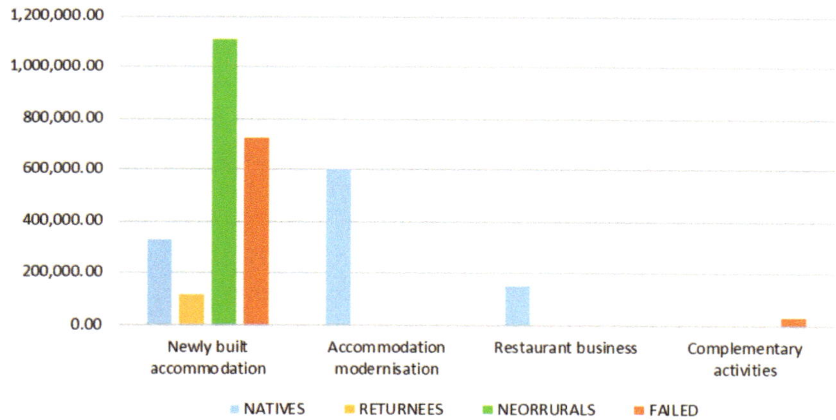

GRAPH 3.2 Investments by activities according to the origin of the promoter.
Source: Own elaboration.

3.3.3 INVESTMENTS AIMED AT TOURISM AND RECOVERY OF THE HERITAGE

3.3.3.1 TRANSVERSAL ACTIONS OF RURAL TOURISM PROMOTION

Within the promotion of rural tourism, almost half of the actions and 15.5% of the investments were managed by public promoters. The following table classifies these actions into several lines of action and quantifies their investment (Table 3.6).

TABLE 3.6 Classification of Public Projects.

Typology of the projects	Number of projects	Investment
Tourism promotion material and campaigns	16	323,140.01
Attending fairs and conferences	7	32,055.77
Tourist information and assistance centers	4	245,004.35
Projects and activities for tourism revitalization	4	33,996.56
Signposting hiking trails	1	23,372.68
Total	**32**	**657,569.37**

Source: Own elaboration.

As seen, this type of investment is not aimed at achieving an immediate economic benefit, but rather aims at promoting tourism in the region from a number of perspectives: advertising and promotion campaigns, attending tourism fairs, creation of infrastructures for the provision of information services and tourist assistance, valorization of tourist resources, etc.

Although the logic of these actions is very different from that followed by private projects, there is no doubt that indirectly, this type of investment complements those of private promoters, even to their benefit, and is fundamental within the strategy of tourism development undertaken by the region of La Vera.

3.3.3.2 COMPLEMENTARY INVESTMENTS OF RECOVERY OF THE HERITAGE

Apart from the investments included in the measure to promote tourism, the development strategies are undertaken by the regions that manage the programs under study also include a series of projects that receive mainly public funding, complement the investments addressed within productive measures and, especially, those related to the promotion of tourism. Under the definition of *"Recovery and conservation of the environment"*, Table 3.1 includes this type of investment. They are very diverse actions ranging from social infrastructures to immaterial actions aimed at recovering the cultural heritage and popular traditions (Graph 3.3).

The analysis of the actions included in the previous graph provides a more complete vision of the strategy proposed for the promotion of the tourism sector. It is noteworthy that almost a tenth of the resources used are directed at actions aimed at the recovery and/or enhancement of several tourist resources: bathing areas in natural gorges, recreational areas, lookouts, and even tourist information points.

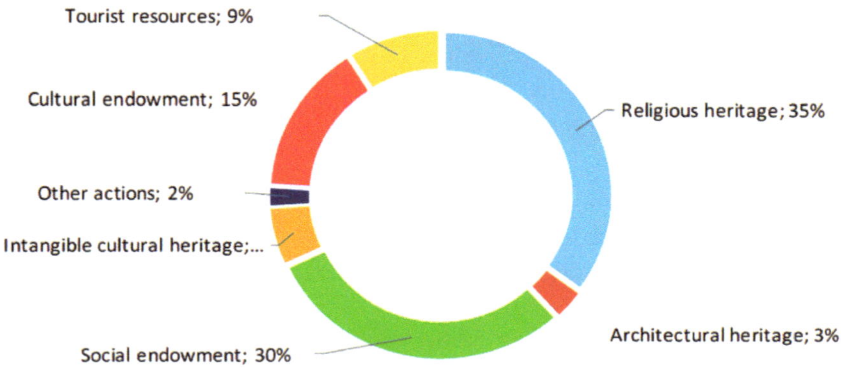

GRAPH 3.3 "Non-productive" investments for environmental recovery.
Source: Own elaboration.

Those actions related to the recovery of religious and architectural heritage can be highlighted among those other complementary actions. The resources used in these two concepts are around 40% of the total investment of this measure. These two lines of action are materialized in very heterogeneous actions: while the recovery of religious heritage involves the restoration of religious images, organs, altarpieces, or various rehabilitation actions in hermitages and churches; investments aimed at recovering the architectural heritage involve the recovery of fountains, washing places, castle lighting, etc. There is no doubt that even though they are not specifically aimed at the promotion of tourism, these investments bring value to the region's heritage and convert many of its elements into tourist resources. This demonstrates the close relationship between the recovery of the heritage and the promotion of tourism in rural areas.

The third element that deserves comment is the projects of cultural promotion and recovery of festivals and popular traditions. These actions may be aimed at the local population and their main purpose is to strengthen the identity of the population with its territory, but—not for that reason—should the interest that the consolidation of many of these actions has for the tourism promotion strategy of the region be underestimated.

A marked local character prevails over the rest of the projects executed under this measure. These are endowments and infrastructures aimed at the local population: cultural centers, classrooms for training, nurseries, care centers for the elderly, etc.

3.4 CONCLUSIONS

Due to its characteristics, the choice of La Vera as a field of study is justified given that this region is an ideal framework in which to implement rural development policies and especially those aimed at promoting rural tourism.

The research shows that investments aimed at the promotion of tourism are mainly concentrated on the creation and modernization of accommodation. These projects account for four-fifths of the resources of the aforementioned measure and reduce the support of the program for other types of activities such as restaurant businesses or those aimed at organizing complementary activities for tourists. There is also the paradox that one-third of the investments in the creation of new rural accommodation are unsuccessful. These conclusions can be of great interest to other rural spaces that face development strategies focused on the promotion of their tourism sector.

The involvement of the local population in the implementation of tourism projects remains at levels very similar to those of the promoters of neorrural origin, although there are significant differences between the projects in which they are involved. While neorrurals only invest in the creation of new rural accommodation, natives focus on the modernization of accommodation and restaurant businesses. There is no unsuccessful project within these last two activities.

The support of public funds for strategies to promote tourism in rural areas is essential; not only for co-financing private investments, but also for complementing these with other public actions. This work shows that part of these investments has an exclusive tourism component (promotion campaigns and tourist assistance infrastructures), but also another patrimonial component aimed at the valorization of various elements of rural heritage that can constitute valuable tourist resources.

KEYWORDS

- **case study**
- **cultural heritage**
- **tourism development strategy**
- **economic and social cohesion**

REFERENCES

Arnalte, E.; Oliveira, F. Producción Agraria, Gestión Ambiental y Transición Rural. Tres dimensiones de la cuestión rural en la Península Ibérica. In *Cultura, Inovação e Território. O Agroalimentar e o Rural*; Moreno, L., Sánchez, Mª., Simões, O., Coord., Eds.; Sociedade Portuguesa de Estudos Rurais: Lisboa, 2009; pp. 1–15.

Campesino, A.; Sánchez, M.; Rengifo, J. I. Extremadura. In *La actividad turística española en 2009*; López, D., Del Valle, E., Dir., Eds.; Ramón Areces: Madrid, 2010, pp. 437–450.

Cebrián, A. Génesis, método y territorio del desarrollo rural con enfoque local. *Papeles de Geografía* **2003,** *38*, 61–76.

Coller, X. *Estudio de casos*; Ed. Centro de Investigaciones Sociológicas: Madrid, 2000.

European Comission. *Perspectives for the Common Agricultural Policy*. COM (85) 333. Brussels, 1985.

European Comission. *The Future of the Rural World*. COM (88) 501 final. Supplement 4/88; Brussels, 1988.

European Comission. *Communication Inviting States to Submit Proposals Responding to a Community Initiative for Rural Development* (91/C 73/14); DOCE, March 19, 1991.

European Comission. *Communication Inviting States to Submit a Request for Assistance within a Community Initiative for Rural Development* (94/C 180/12); DOCE, July 1, 1994.

European Comission. Agenda 2000. *For a Stronger and Wider Union*. COM (97) 2000; Luxembourg, 1997.

European Comission. *Communication to the Member States, Laying Down Guidelines for the Community Initiative for Rural Development (Leader +)* (2000/C 139/05). DOCE, May 18, 2000.

Garrido, F.; Moyano, E. Capital social y desarrollo en zonas rurales. Un análisis de los programas Leader II y Proder en Andalucía. *Rev. Int. Sociol.* **2002,** *33*, 67–96.

González, J. El método Leader: Un instrumento territorial para un desarrollo rural sostenible. El caso de Extremadura. In *Desarrollo Rural de Base Territorial: Extremadura (España)*; González, J., Dir., Ed.; MAPA and Consejería de Desarrollo Rural de la Junta de Extremadura: Badajoz, 2006; pp. 13–90.

Guiberteau, A. Fortalezas y debilidades del modelo de desarrollo rural por los actores locales. In *Nuevos Horizontes en el desarrollo rural*; Márquez, D., Coord., Ed.; AKAL: Madrid, 2002, 87–104.

Ivars, J. Turismo y espacios rurales: concepto, filosofía y realidades. *Rev. Invest. Geogr.* **2000,** *23*, 59–88.
Mejías, F. Programas comunitarios de desarrollo rural. Aplicación e impacto en Extremadura. In *Desarrollo Rural de Base Territorial: Extremadura (España)*; González, J., Dir., Ed.; MAPA and Consejería de Desarrollo Rural Junta de Extremadura: Badajoz, 2006, 171–219.
Millán, M. La diversificación en el medio rural como factor de desarrollo. *Papeles de Geografía* **2002,** *36*, 223–238.
Ministry of Agriculture, Fisheries and Food. Programa Nacional Proder I. Madrid, 1996.
Ministry of Agriculture, Fisheries and Food. RD 2/2002 Which Regulates the "Leader +" Initiative and the Rural Development Programs (PRODER). BOE, n° 11, January 12, 2002.
Navarro, F.; Woods, M.; Cejudo, E. The Leader Iniciative has been a Victim of Its Own Success. The Decline of the Bottom-up Approach in Rural Development Programmes. The Cases of Wales and Andalusia. *Sociología Ruralis*, **2016,** *56* (2), 270–288.
Nieto A.; Cárdenas, G. Análisis del método Leader (2007–2013) en Extremadura mediante técnicas SIG y análisis multivariado. *Cuadernos Geográficos* **2017,** *56* (1), 148–171.
Nieto, A.; Gurría, J. L. El modelo rural y el impacto de los programas Leader y Proder en Extremadura (Propuesta metodológica). *Scripta Nova* **2010,** *340*, 1–25.
Pisani, E. *Pour una Agriculture Marchande et Ménagére*; Editions l'Aube, 1994.
Pulido, J.; Cárdenas, P. El turismo rural en España. Orientaciones específicas para una tipología aún en desarrollo. *Boletín de la Asociación de Geógrafos Españoles* **2011,** *56*, 155–176.
Ramos, E.; Garrido, D. Estrategias de desarrollo rural territorial basadas en las especificidades rurales. El caso de la marca Calidad Rural en España. *Estudios Regionales* **2014,** *100*, 101–129.
Ruíz, P. Perspectivas y nuevas orientaciones del turismo rural. In *Hacia un nuevo sistema rural*; Ramos, E., Cruz, J., Coord., Ed., MAPA: Madrid, 1995, 495–515.
Zapatero, J.; Sánchez, M. J. Instrumentos específicos para el desarrollo rural integrado: La Iniciativa Comunitaria LEADER y el Programa Operativo PRODER. *Polígonos* **1999,** *8*, 21–38.

CHAPTER 4

CULTURAL AND ECOSYSTEM SERVICES FROM A NATURAL HERITAGE SITE: A CASE OF SUNDARBAN

SHAFIQUL ISLAM[1*] and UMMUL FATIHA BINTE SHAFIQ[2]

[1]*Center for Sustainable Development, University of Liberal Arts Bangladesh*

[2]*Department of Environmental Science & Engineering, Jatiya Kabi Kazi Nazrul Islam University, Bangladesh*

[*]*Corresponding author. E-mail: shafiqul.islam@ulab.edu.bd*

ABSTRACT

Sundarban plays an important role in ecotourism and livelihood development as the natural heritage site. Both domestic and international tourists visit Sundarban for recreation and learning. International visitors come to see the World Heritage Site Sundarban crossing the boundaries. There is no boundary for heritage and cultural tourism. Revenue comes from tourism and indirect support services contribute to the local and national economy through natural heritage tourism. Direct and indirect employment opportunities have been created for the community. Boat riding, fishing, local food, accommodation, local transport, handicrafts, and guide services are the important field of employment. At the natural site like Sundarban, people can present their culture (folk song, occasional festivals) and heritage (temples) to domestic and international tourists. Students and researchers come to the site for education and research for gaining new insights into it. Sundarban also teaches people about biodiversity, wildlife habitats, protection and regulation services, and how to living with nature. It was understood that human endurance pursuit is connected with the attraction of the adjoining environment. Nature has been departed due to human activities and so-called progress, but the heritage site put them together for the conservation of their culture and identity. Heritage tourism is the safest effort for the protection of frail flora and fauna as well as makes better engagement opportunities to the community people. The Sundarban resolves many problems for the community including their livelihoods, natural resources conservation, cultures, values, and the spirit of the local community. Cultural ecosystem services and traditional ecosystem knowledge can connect the local community and mangroves for the sustainability of culture, heritage, and livelihood. There is no boundary of natural heritage tourism from international perspectives. Local culture and history is the main concern of natural heritage.

4.1 INTRODUCTION

The Sundarban is the biggest mangrove forest in the world covers 10,000 km^2 land and water. It also covers a total of 75 compartments of which 55 compartments in Bangladesh and the remaining part in India. It occupies on the delta of the Ganges, the Brahmaputra, and the Meghna rivers on the Bay of Bengal. The geographical location of Sundarbans Reserve Forest between latitude 21° 27′ 30″ and 22° 30′ 00″ North and longitude 89° 02′ 00″ and 90° 00′ 00″ East. The land of Sundarban composed of exposed sandbars (70%) and water bodies (30%). The annual rainfall of the Sundarban ranges from 1500 to 2000 mm with the estuarine water temperature 20–35°C (Rahman and Asaduzzaman, 2010; Sunny, 2017). It locates at the coastline (low), with tidal water reaching up to 110 km inland and extended mangrove (Spalding et al., 2010). More than 5 million people are living in the Sundarban peripheral regions from India and Bangladesh for their livelihoods (Neogi et al., 2017). In 1997, the UNESCO has recognized and declared the Sundarban as the World Heritage Site (Islam and Islam, 2003). The Sundarban contains a wide range of fauna and flora. It was reported that Sundarban is the harbor of 70 plant species from 34 species (Heining, 1892) and 1136 wildlife species (Forest Department, 2018). It is quite impossible to maintain the Sundarban by "command and control" and tourism-based interventions only (Chakraborty et al., 2018). Rather cultural services, several factors of socioeconomic and eco-political bottlenecks including poverty, and breaching of laws and corruption prevail at many levels (Rahman et al., 2010). Different services are provided by this natural heritage such as provisioning direct services (timber, fuel, housing materials, and nontimber forest products), cultural services (tourism), and protection services like cyclone, storm, tidal surges, and flood (Uddin et al., 2013). Almost 3.5 million people are getting all sorts of services from Sundarban Mangrove (Giri et al., 2007). Services and natural heritage are exposed by natural factors as well as human-induced factors. Sea level rise (Gilman et al., 2008) and stress to biodiversity (Luetz, 2008) are important among natural factors. According to the report of FAO (2010), 112 countries have few classified mangrove areas covering 15.6 million hectares and 100 countries have no mangroves out of 212 countries. Biswas et al. (2007) reported that Sundarban provided a variety of services including storing of sedimentation for land acclimatization, human lives protection from storm, cyclones, tidal surges, and inundation, livelihood support for cash economy, water and waste recycling, carbon cycling and oxygen production. Heritage tourism provided important services to the people crossing boundaries for national and international tourists from economic perspectives. Traditional ecosystem knowledge (TEK) is very essential for the management of mangroves (Queiroz et al., 2017) at present. The fisheries in the Sundarban have become commercialized (Kroodsma et al., 2018) as well. In Bangladesh the largest mangroves Sundarban plays a key role in the heritage tourism. It attracts tourists apart from Bangladesh due to its recognition as the World Heritage Site and diversified characteristics. Nilkomol, Kachikhali, Kotka, Dublar Char, Shekhertek, and Bari temple are the most attractive tourist spots in Sundarban. Community engagement is very important in heritage tourism to demonstrate their culture, tradition, and role (Islam, 2019). Manyara and Jones (2007) mentioned that community tourism has a direct economic impact on families, socioeconomic, and diversified livelihoods.

4.1.1 CONCEPTUAL FRAMEWORK OF THE STUDY

The study is based on a typical conceptual framework and the details framework is given in Figure 4.1. It is essentially important to understand the importance of services and livelihood support provided by natural heritage like Sundarban. Eventually, it is important to know how natural heritage Sundarban ecosystem services are contributing for the well-being of the people as well for the country. The study assessed the contribution of

Sundarban to the cultural ecosystem services (CES) (tourism and regulatory services) and livelihood. The study also investigates how Sundarban provides cultural services (tourism) for the sustenance of livelihoods.

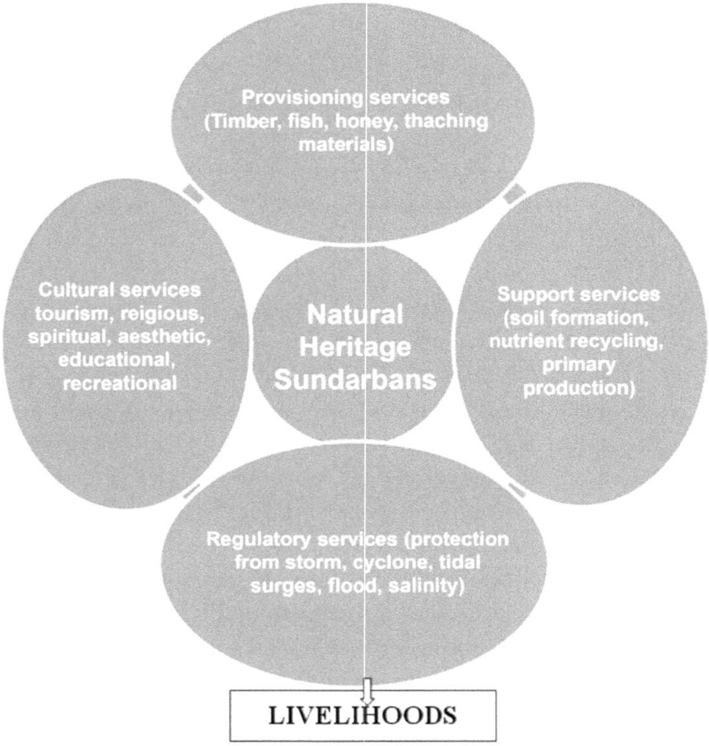

FIGURE 4.1 Conceptual framework for the study.

Source: Adapted from De Groot et al. (2002).

4.1.2 RESEARCH QUESTIONS

The following research questions have been deployed for this study:

- What are services provided by natural heritage site (Sundarban)?
- How ecosystem services support livelihoods?
- What are the cultural (tourism) services provided by natural Sundarban?
- At what extent and how heritage tourism provide livelihood support to the community people?

4.1.3 OBJECTIVE

The aim of this study is to evaluate how natural heritage provides services and livelihood support for community development. The study addressed the following specific objectives:

- To elicit services that are provided by Sundarban as a natural heritage site
- Know the site-specific cultural services that derived from Sundarban
- Make linkages with cultural services and livelihood sustainability
- To find out problems related to heritage tourism in Sundarban
- To recommend solutions overcome the problems

4.2 METHODOLOGY

The study was conducted in the Sundarban focusing tourist point Nilkomol, Kotka, Koromjol, Kachikhali, and Dublar Char (Fig. 4.2), regulating authority (Forest Department (FD) and Police station), and community. Five Focus Group Discussions (FGD) were carried out for this study. Five Key Informants' Interviews and secondary information were used for this study. The qualitative and quantitative data were deployed for this study. FGD with the concerned people and small households (community people) were carried out for this study. The monetary valuation of cultural services (heritage tourism) has been collected from the Department of Forest (secondary data). Both domestic and international tourists have been considered in this calculation.

4.3 RESULTS AND DISCUSSIONS

4.3.1 THE WORLD HERITAGE SUNDARBAN AS THE PROTECTED AREA

The World Heritage Site Sundarban is a great place for tourism and coastal protected areas covering three districts such as Bagerhat, Khulna, and Satkhira. The forms of the protected areas are wildlife sanctuaries, fish sanctuaries, Ramsar site, the World Heritage Site, and fish sanctuaries (Table 4.1). The tourists come to visit such places with the permission from the concerned department (Forest Department).

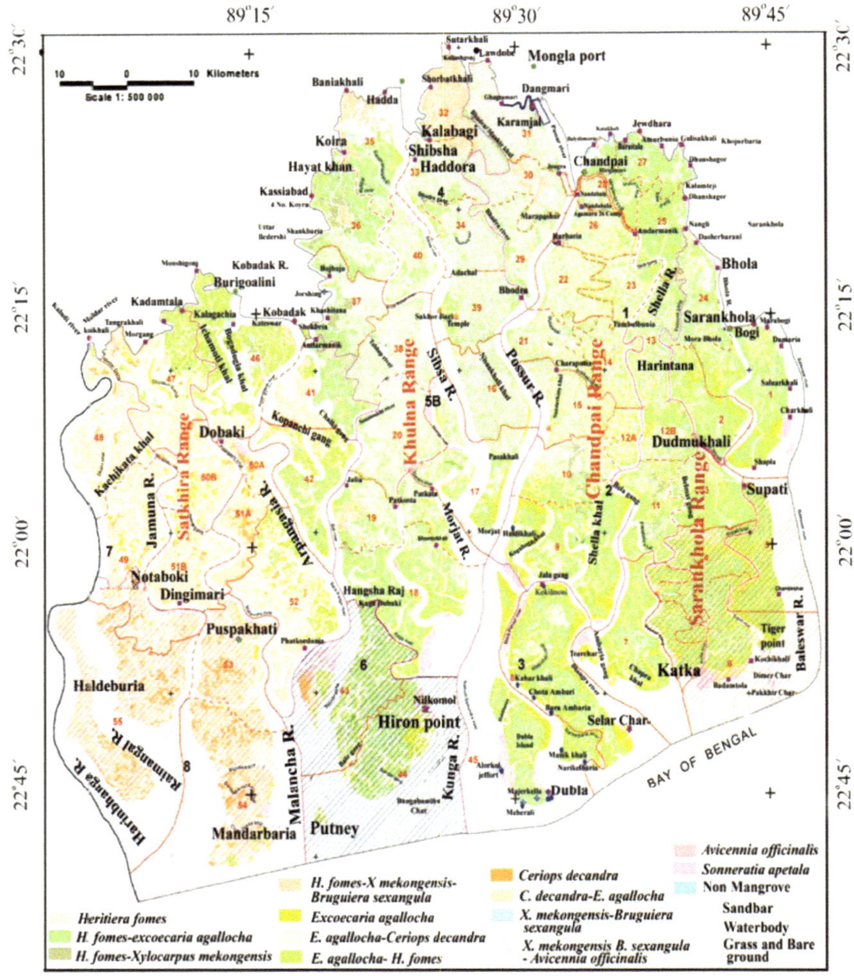

FIGURE 4.2 Location map of the study area.

4.3.2 IMPORTANT PLACES OF THE NATURAL HERITAGE SITE

Several important places were mentioned by the respondents and visitors as the unique place for the natural heritage site. Katka-Kachikhali, Dubla Island, Nill Komol or Hiron Point, Mandarbaria, and Karamjal are the important places for the site.

TABLE 4.1 Protected Areas of Sundarbans.

Category	Name	Area (ha)	Location
Wildlife sanctuaries	Sundarbans East	31,227	Bagerhat
	Sundarbans West	71,502	Satkhira
	Sundarbans South	36,970	Khulna
Ramsar site	The Sundarbans	601,700	Bagerhat, Khulna, and Satkhira
World Heritage Site	Wildlife sanctuaries of the Sundarbans		Bagerhat, Khulna, and Satkhira
Fish sanctuaries			Bagerhat and Khulna

Source: Adapted from Islam (2004).

4.3.2.1 KATKA/KACHIKHALI

Katka/Kachikhali is situated to the vicinity of the Bay of Bengal and exhibits its beautiful *khals*, flora and fauna. It is one of the important heritage sites of Sunderban. The site provides safaris and the spot to look tigers and watching beautiful birds. The site is the place of numerous rare and grand wild fauna. It was reported by the respondents that the *khals* of the site offer visibility of deer (*Cervus unicolor*), monkeys (*Macaca mulata*), Estuarine Crocodile (*Crocodilus porosus*), masked fin-foot, Gangetic Dolphin (*Platinista gangetica*) monitor lizard (*Varanus salvator*), snakes (*Python bivittatus*), and often a tiger (*Panthera tigris*) during crossing a *khal*. There is an observation tower through which tourists can see a large number of different types of deer and sometimes tiger can be found. This place is also known as the hot spot of the Sundarban.

4.3.2.2 DUBLA ISLAND

Dubla Island is very much famous for dry fish. It is the only accessible place for fishing and processing centers especially for dry fish during winter. *Rashmela* is a religious festival that takes place each and every year in the month of November during the full moon. A group of visitors comes this time in Dubla Island as reported by the FGD participants.

4.3.2.3 NILL KOMOL/HIRON POINT

The beautiful place of the Sundarban is Nilkomol or Hiron point. This place attracts the tourists from home and abroad to see the unique beauty of the site. It has been erected as the World Heritage Site.

4.3.2.4 MANDARBARIA

It is the place on the southwest side of the Sundarban. This place is very isolated and quiet in nature. Tourists like this place for the exclusive wilderness. The nature lover and silence people love this place more than other places of the Sundarban.

4.3.2.5 KARAMJAL

The place is the key entry point to the Sundarban. Visitors/tourists come here and can move through the Sundarban Reserved Forest. Few wild species have been kept here for captive breeding and most of the visitors like it. Visitors can find mangrove arboretum in Karamjal. The place should be developed as a center for interpretation.

4.3.3 SERVICES OF THE WORLD HERITAGE SITE SUNDARBANS

Heritage site the Sundarban offers multiple services and credits for the country and nations. It is evident that a heritage site is an excellent place for tourism that conserves the local culture and tradition for livelihood development. The services provided by Sundarban are direct (provisioning) services, ecosystem services, protection or regulatory services, cultural (tourism) services, and support services for livelihood sustenance. Table 4.2 shows the different services provided by the Sundarban. It was reported that the World Heritage Site Sundarban provided revenue worth of USD 879,181 in 2013 (Forest Department, 2018).

TABLE 4.2 Services Provided by Sundarbans.

Direct services	Ecosystem services	Heritage services	Protection services	Support services
Timber (*Heritera fomes; Sonneratia apetala)*	Pollination	Aesthetic	Cyclone and storm,	Biodiversity
Fuel (*Ceriops decandra*)	Soil formation	Ecotourism	Tidal surges	Habitats for animals
Food (*Nypa fruits and nolen gur)*	Water purification	Heritage and identity	Flood	Habitats for plants
Thatching materials (*Nypa fruiticans*)	Pollution abatement	Local culture	Soil erosion	Soil stabilization
Raw materials for paper (*Excoecaria agallocha*)	Carbon sequestration	Recreation	Nutrient recycling	Breeding ground
Housing materials	Climate regulation	Education and research	Seeds germination/seedling anchorage	Protection from sedimentation
Fishes	Food chain	Holy place	Protection from salt intrusions	Biomass formation
Honey	Social cohesion	National forest of Bangladesh	Crop pest regulation	Employment
Medicinal plant	Existence	Inspiration for nature lovers		
Mollusk		Tradition of local livelihoods		
Crab		Spiritual		
Wild animal (Deer, snake, crocodile, dolphin)		Sense of place		
Fooder				

Source: Adapted from Forest Department (2018).

4.3.3.1 DIRECT SERVICES

The Sundarban makes a direct contribution of revenues through provisioning services for timber fuel, honey, fish, thatching materials for livelihood support, and the economic development of the country. Well managed selective method of extraction forest goods and nontimber forest products are the key tools for the sustainability of mangrove forests.

4.3.3.2 ECOSYSTEM SERVICES

A range of ecosystem services provided by the world heritage Sundarban include water recycling, oxygen production, carbon storing and recycling, waste recycling, pollination services, and soil formation. The safeguarding of water quality, balancing salinity, and deposit ruling are important functions of the Sundarban (Mohiddin et al., 2015). Carbon sequestration is one of the important ecosystem services provided by the Sundarban. It was assumed that the carbon reserve of Sundarban possibly will be around 112 million tons amounting to 0.56–1.68 million each year. Hossain (2014) reported that it provides protection from the storm and coastal erosion and ecosystem services with USD 273–714 million per year. The World Heritage Site also provides CES to maintain coastal ecosystem services and conservation from nonutilitarian perspectives (Chaktaborty et al., 2018). CES render numerous benefits to humans such as aesthetic, spiritual, psychological, and nonmaterials (Millennium Ecosystem Assessment, 2005). Till now, the services of CES are not well recognized when compared to other services including provisioning, regulating, and supporting services (Rodrigues et al., 2017).

4.3.3.3 CULTURAL (HERITAGE TOURISM) SERVICES

It is well known to all that the Sundarban is a tourism destination and provides natural attractions to the visitors including the Royal Bengal Tiger, estuarine crocodile, mangrove vegetation, and different avian. Diverse heritage services are provided by the Sundarban through domestic and international tourism activities. These services include ecotourism, World Heritage Site, national reserved forest in Bangladesh, a place of tradition (*rashmela*), country branding value, demonstrate local culture, inspiration source, education and research value, and traditional livelihoods. The World Heritage Site Sundarban provides a home of cultural services including education, tourism, ethnic festivals, and adoration by local people. It is not the only motto of people to go to a place (historic) for sightseeing. The tourists are curious to gain knowledge and experience. The tourists visit the natural heritage site to understand the importance of a place and history of many cultures. Total USD 1,056,036 revenue was collected during the period of 2003–2017. It was found that the highest number of tourists visited Sundarban in the year 2016 (Table 4.3). In the year 2017, total visitors were 191,990 individuals of which 1,888,965 were domestic visitors and 3025 were international visitors. Corresponding figures, the fewer numbers of visitors found in the year of 2009 and 2015 due to political instability. The diversified characteristics of Sundarban attract the tourist from local to globe and play an important role in economic contribution. The most attractive tourist places in Sundarban are Kotka, Kachikhali, Nilkomol, Nilkomol, Shekhetek, and Bari Temples. The income from tourism is also contributed to the economy either directly and indirectly. These places are very much vulnerable to natural calamities especially sea level rising that will affect tourism in the world natural heritage site (Sundarban). According to Bangladesh Parjatan Corporation, 5 out of 18 spots are in the coastal areas of Bangladesh. Two important spots are located in the Sundarban out of five spots in the coasts. Several festivals and special events are celebrated within Sundarban and its adjoining places that can attract tourists as mentioned by the FGD participants. River cruising, fishing, beach relaxation, walking, bird, dolphin and wildlife watching, jungle trails, and boat trips are the most important activities for the tourists. Such types of activities attract both local and international communities to come and experience. Visitors come to the Sundarban for recreation and enjoyment. Many students come here for learning and field exposure. They enjoy the wild animals (deer, monkey, wild boars, different kinds of birds, and crocodile. They mostly come to see plants like Sundri, Keora, Baen, and Goal pata. Pneumetaphore is the key attraction for them.

4.3.3.3.1 Entry Points and Visitors

Four formal entry points have been reported by the Forest Department. These entry points are Chandpai, Dacope, Munshigonj, and Sharankhola.

The maximum numbers of visitors entered the site through Chandpai followed by Munshiganj, Sharankhola, and Dacope, respectively (Fig. 4.3). The number of visitors, revenue, and entry points were considered during the period of 2012–2017. It was found that the highest numbers of international tourists entered Sundarban through Chandpai followed by Sharankhola, Munshiganj, and Dacope. Respondents mentioned that communication is easy through Chandpai.

TABLE 4.3 Status of Visitors and Revenue Generated from Tourism.

Year	Total number of visitors	Domestic visitors	International visitors	Total revenue (USD)
2003	59,169	58,737	432	19,298
2004	50,093	49,366	727	29,071
2005	48,828	46,887	1941	33,729
2006	71,202	69,078	2124	29,314
2007	94,214	92,632	1582	48,637
2008	96,002	94,745	1257	41,743
2009	85,188	83,709	1479	39,371
2010	99,877	97,721	1706	63,115
2011	119,256	117,100	2156	82,398
2012	39,350	37,493	1857	47,803
2013	105,968	102,905	3063	115,656
2014	137,129	133,725	3404	119,694
2015	88,653	85,941	2712	94,587
2016	191,990	188,965	3025	188,947
2017	129,853	128,345	1508	102,673
Total	1,416,772	1,387,349	28,973	1,056,036

Source: Adapted from Forest Department (2018).

4.3.3.3.2 *Purpose of Visit*

Respondents mentioned about the nine purposes of the visitors to come in the Sundarban. Education, research, excursion, see wildlife, enjoy beauty, see plant species, riding boat, recreation, and enjoy holidays are the purpose of visitors to come Sundarban. Most of the visitors come for enjoyment and excursion (Fig. 4.4).

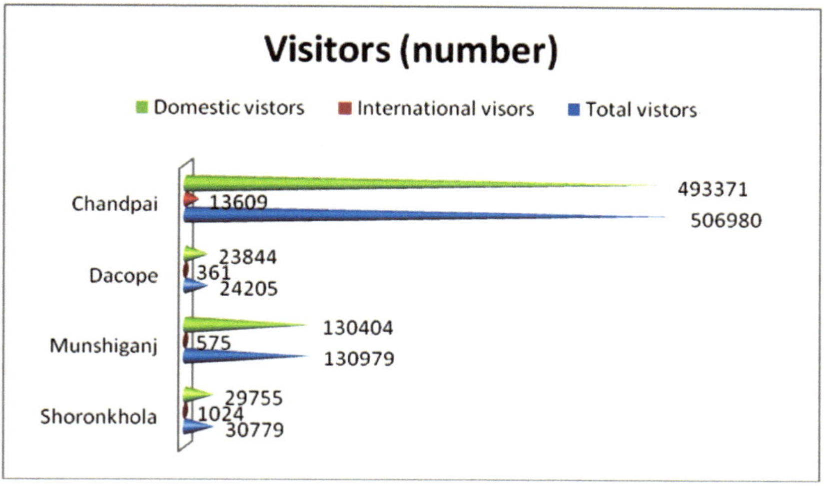

FIGURE 4.3 Entry points and number of visitors.

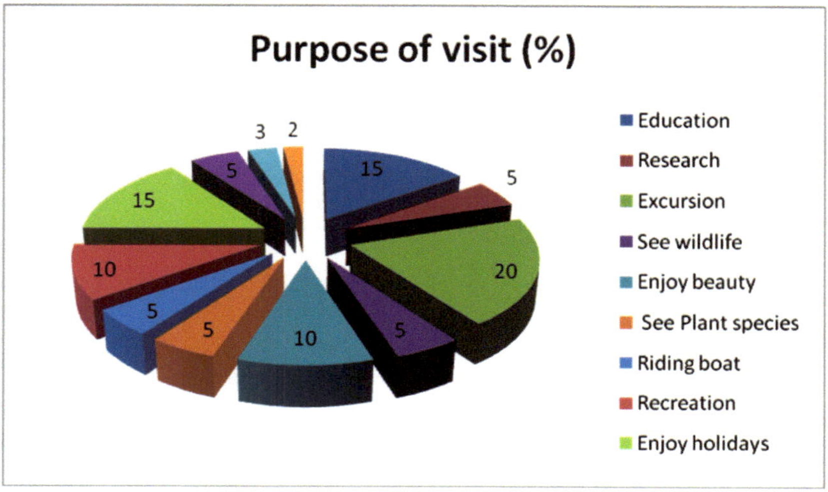

FIGURE 4.4 Purpose of visits.

4.3.3.3.3 *Visitors/Tourists*

The satisfaction level of tourists depends on several factors. The factors including the tourists' number who are visiting the place, their comments, continue to stay as well their aspiration to make replicate visits. Usually, the visitor makes comments just after visiting the place. Regarding the beauty and uniqueness of the sites, the visitors are satisfied but in relation to the products, the visitors are skeptical. They reported that the quality and readiness of the products are limited. It is found from Table 4.2 that the number of tourists visiting Sundarban is increasing with time. Both national and international tourists are not pleased since tiger sighting in the wild at the heritage site is exceptional. Few of them did not want to visit yet again to Sundarban. In the year 2015, the beach city Coxsbazar did not see a sizeable inflow of tourists, owing to the ongoing political tension that has claimed many lives and made traveling less desirable.

4.3.3.4 PROTECTION OR REGULATORY SERVICES

As the natural heritage site, the Sundarban has been paid attention in the world for conservation and protection. It provides lots of protection and regulatory services such as protection from cyclones, tidal surges, flood, erosion control, and carbon storage or sequestration. The Sundarban provides excellent regulation services. It saves lives and assets of community people adjacent to the Sundarban. The site also acts as a safeguard or barrier to the storms and cyclones. Roots and trees of mangrove provide protection from tidal surges, storms, seepage, and salinity intrusion (Walters et al., 2008).

4.3.3.5 LIVELIHOOD SUPPORT

The heritage Sundarban has been provided livelihood support for the forest dwellers and neighboring forest people through provisioning services, ecosystem services, regulatory services, and cultural (tourism) services. In case of heritage tourism (cultural services), the community people get employment through several activities. The employment services have been generated through guide services, preparation of food, fair (*rashmela*), pottery, folk songs, folk medicine, local transport (boat, van, and rickshaw), support services, staff in the hotel, restaurant, and handicrafts preparation. Sundarban is an excellent site for livelihood development. It occupies a lot of people for diversified livelihoods opportunities including fishing, collection of fuel, timbers, honey, *goalpata,* and crabs. It was reported by the FGD participants that around 55% of respondents earned their 70%–80% income from provisioning services. It was reported that more than 3.5 million people depend on Sundarban directly for their livelihoods (Mozumder et al., 2018). Engagement of local community for facilitating ecotourism might be contributed local economy significantly (Mukherjee et al., 2014). The respondents opined that there is an urgent need of ecotourism activities for the conservation of the heritage site. Fishing is the most common and important income-generating activity that provides livelihood support. The people deployed in this profession through the local trader or by the moneylender. The condition is that the people sell their harvested fish to the local leader or money lender. Mainly men were involved in fishing. Both men and women involved in fry collection. The second option for livelihood is fuelwood collection. They also lead their livelihood through crab collection, *goal pata,* and honey collection.

4.3.4 CONTRIBUTION OF NATURAL HERITAGE TO SUSTAINABLE DEVELOPMENT

The Sundarban is a great natural heritage site in Bangladesh. It is also known as the World Heritage Site. It contributed toward the pillars of sustainable development. The participants from FGD have been provided valuable insights about the contribution of Sundarban to sustainable development. Sustainable matrix exercise has been performed for this determination. It was mentioned by the respondents that protection services are the topmost priority followed by ecosystem and cultural services, provisioning services, and support services, respectively (Table 4.4).

4.3.5 PROBLEMS WITH THE WORLD HERITAGE SITE SUNDARBAN

Several challenges hinder the opportunities of heritage tourism. Both natural and human-induced factors are affecting the natural heritage site for sustainability and conservation. Human-induced factors include illicit cutting of trees, encroachment of forest lands, diversion of fresh water, oil spill, navigation, shrimp cultivation, use of chemicals (agrochemicals), pollution, and poaching. It has been identified that the plant growing stock reduces from 296 to 144 during 1959–1996 in Sundarban (Giri et al., 2015). A similar observation has been made by Azizul and Paul (2015) putting emphasis on

TABLE 4.4 Sustainability Matrix and Contribution of Sundarban to Sustainable Development Goals.

Services	Major products/services	Contribution of services to SD pillars			Total score	Rank	Contribution to SDG 14 and 15
		Economic	Environment	Social			
Provisioning Services	• Timber • Fuel • Honey • Fish • Housing materials	5	2	5	12	3	14.1, 14.2,14.4, 15.1, 15.2, 15.5, 15.9
Ecosystem services	• Existence • Water recycling • Pollination • Climate regulation	4	5	5	14	2	14.B, 15.5, 15.9, 15.C
Cultural (Heritage) services	• Ecotourism • Education • Recreation • Aesthetic • Research • Identity • Spiritual	4	5	5	14	2	14.2, 14.4,14.5, 15.2, 15.5, 15.7, 15.9

TABLE 4.4 *(Continued)*

Services	Major products/services	Contribution of services to SD pillars			Total score	Rank	Contribution to SDG 14 and 15
		Economic	Environment	Social			
Protection services	• Cyclone • Storm • Flood • Tidal surges • Insect pests	5	5	5	15	1	14.B,15.5, 15.9, 15.C
Support services	• Biodiversity • Habitats • Breeding • Soil stabilization • Employment	3	5	4	12	3	14.4,14.B, 15.1, 15.9

Source: Authors and websites: https://sustainabledevelopment.un.org/sdg14, https://sustainabledevelopment.un.org/sdg15.

Score: Very important = 5, Important = 4, Moderately important = 3, Less important = 2, Not important = 1.

the construction of Farakka barrage in the upstream, coal-fired power plants construction, industrial pollutants, heavy metal deposition, and execution of the law. Nature also is impacting the site through cyclonic storms and tidal surges by eroding soil, destroying human settlements, uprooting trees, breaking stem, and branches. Several cyclones (Sidr in 2007, Aila in 2009, and Mahasen in 2013) caused huge damage to the mangroves and human settlement in the ruthless areas of the Sundarban (Aziz and Paul, 2015). It has been identified that a significant decrease (1.1%) of forest lands of total Sundarban during the period of 1970–2000 (Giri et al., 2015). Sundarban is reducing at an alarming rate. It was found that *H. fomes* decrease 28.75% during 1989–2010 and total tree cover reduced 3% during the same period of time (Aziz and Paul, 2015). Infrastructure (road and standard hotel) and tourism products are not readily available in this site. Security is an important issue especially for the international visitors. The major challenges of managing Sundarban ecosystems are conservation and protection. Similarly, Chakraborty (2018) reported about axial coding and selective coding which is concerned with the mangrove sustainability. The quality of the tourist products in the Sundarban is not up to the standard. The engagement of local people in tourism activities is very minimal. Income generation for the community people from tourism or cultural services is very poor. The site is also lacking from the trained and qualified tourist guide. Eventually, the quality and services for the tourists are not available. The community people do not have hands-on experience dealing with tourists. Seasonality is also another problem for the tourist to make visit Sundarban. Traditional resource harvesters of the Sundarban are very much concerned about conservation, but TEK falls into a vicious cycle that is giving birth to huge corruption, liability, and bribe. The system would allow the nontraditional resource harvesters by enduring illegal activities such as poisoning upstream for fishing, clear-felling, and illicit cutting of wood.

4.4 CONCLUSION AND RECOMMENDATIONS

Plants stock has been drastically reduced during the period of 1959–1996 and it is also expected that decreased numbers of plants may be lesser in the near future. TEK is very much important for the conservation of Sundarban mangroves. TEK is threatened by several drivers including degradation of resource harvesters' socioeconomic situation and mangrove degradation (Chakraborty et al., 2018). Chakraborty et al. (2018) suggested for the cultural interactions through selective extraction of resources (timber, fuel, fish, honey, etc.) limited access and metaphors of tigers (protectors) for the maintenance of ecosystems. As the world heritage site, Sundarban can provide the best places and services for tourism. In this connection, the government needs to pay more attention to policy and strategic plans for the conservation of such heritage site. Security is the very first important factor to attract a large number of tourists (national and international) for heritage tourism. A sound environment and a politically stable country are therefore essentially important if we want to attract travelers. Promotion and branding are also very important for heritage tourism. Before leaving the site, the visitors often say that they did not know where to visit and even there were so many things to see and experience here. It is essentially important to promote heritage sites, history, culture, and cuisine properly. The product should be ready for promotion, branding, and experience. Of course, international visitors will not come to a country and visit while risking their lives. The country image is also important to potential international tourists. The Sundarban has ample scope of heritage tourism and to grow popular tourist places in Bangladesh. Actually, Sundarban possesses a rich heritage and diverse culture. It is proven that the tourist loves excitement to enjoy their leisure time for recreation. Sound infrastructure is the basic requirement for tourism destinations. The World Heritage Site Sundarban maintained critical habitats for terrestrial and marine species diversity. International policy for the conservation of natural heritage and World Heritage Site is essentially important and linked with human well-being (Pandey et al., 2016). As the natural heritage site, the Sundarban has ample scope of tourism for the national and international stakeholders.

KEYWORDS

- **cultural ecosystem services**
- **ecosystem services**
- **natural heritage**
- **Sundarban**
- **traditional ecosystem knowledge**
- **recreation**

REFERENCES

Aziz, A.; Paul, A. Bangladesh Sundarbans: Present Status of the Environment and Biota. *Diversity* **2015,** *7* (3), 242–269.

Biswas, S. R.; Choudhury, J. K.; Nishat, A.; Rahman, M. M. Do Invasive Plants Threaten the Sundarbans Mangrove Forest of Bangladesh? *For. Ecol. Manage.* **2007,** *245* (1–3), 1–9.

Chakraborty, A. Challenges for Environmental Sustainability in a Mountain Destination: Insights from the Shiroumadake District of North Japan Alps. *Geo J.* **2018,** 1–21. https://doi.org/10.10708-018-9868-1.

Chakraborty, S.; Saha, S. K.; Selim, S. A. Meanings, Opportunities and Challenges of Cultural Ecosystem Services-based Coastal Management in the Sundarbans Mangroves, Bangladesh: Management Priorities for Achieving SDGs 14 and 15. In *The Environmental Sustainable Development Goals in Bangladesh*; Routledge: London, 2018; pp. 115–131.

De Groot, R. S.; Wilson, M. A.; Boumans, R. M. A Typology for the Classification, Description and Valuation of Ecosystem Functions, Goods and Services. *Ecol. Econ.* **2002,** *41* (3), 393–408.

Food and Agriculture Organization (FAO). Global Forest Resource Assessment 2010. FAO Forestry Paper 189. Food and Agriculture Organization of the United Nations Rome, Italy.

Forest Department. *Sundarbans: The World Heritage Site, Divisional Forest Office*; Sundarbans West Forest Division: Khulna, Bangladesh, 2018.

Gilman, E. L.; Ellison, J.; Duke, N. C.; Field, C. Threats to Mangroves from Climate Change and Adaptation Options: A Review. *Aquat. Bot.* **2008,** *89* (2), 237–250.

Giri, C.; Long, J.; Abbas, S.; Murali, R. M.; Qamer, F. M.; Pengra, B.; Thau, D. Distribution and Dynamics of Mangrove Forests of South Asia. *J. Environ. Manage.* **2015,** *148*, 101–111.

Giri, C.; Pengra, B.; Zhu, Z.; Singh, A.; Tieszen, L. L. Monitoring Mangrove Forest Dynamics of the Sundarbans in Bangladesh and India Using Multi-temporal Satellite Data from 1973 to 2000. *Estuar. Coast. Shelf Sci.* **2007,** *73* (1–2), 91–100.

Heinig, R. L. Working Plan of the Sundarbans Government Forests, Khulna and 24 Parganas Districts, Bengal, 1892.

Hussain, M. Z. Bangladesh Sundarban Delta Vision 2050: A First Step in Its Formulation—Document 2: A Compilation of Background Information. *IUCN, International Union for Conservation of Nature, Bangladesh Country Office*; Dhaka, Bangladesh, 2014; pp. 1–192. https://sustainabledevelopment.un.org/sdg14, https://sustainabledevelopment.un.org/sdg15.

Integrated Resource Management Plans for the Sundarbans; Forest Department, Ministry of Environment and Forests: Dhaka, Bangladesh, vol. 1, 2010; pp. 1–281.

Islam, M. S. Community Development through Promoting Ecotourism in Mangrove Areas of Bangladesh. In *Conservation and Promotion of Heritage Tourism*; IGI Global, 2019; pp. 155–179.

Islam, M. S. N.; Islam, S. M. N. *Sustainable Eco-tourism as a Practical Site Management Policy?: A Case Study on the Sundarbans World Heritage Site in Bangladesh*; AH Development Publishing House, 2003.

Kroodsma, D. A.; Mayorga, J.; Hochberg, T.; Miller, N. A.; Boerder, K.; Ferretti, F.; ... Woods, P. Tracking the Global Footprint of Fisheries. *Science* **2018,** *359* (6378), 904–908.

Luetz, J. Planet Prepare: Preparing Coastal Communities in Asia for Future Catastrophe. Asia Pacific Disaster Report, 2008.

Manyara, G.; Jones, E. Community Based Tourism Enterprises Development in Kenya: An Exploration of their Potentials as Avenues of Poverty Reduction. *J. Sustain. Tour.* **2007,** *15* (6), 628–644. DOI: 10.2167/jost723.0.

Millennium Ecosystem Assessment. *Ecosystem and Human Well-Being: Synthesis*; Island Press: Washington, DC, 2005.

Mohiuddin, K. M.; Alam, M. M.; Arefin, M. T.; Ahmed, I. Assessment of Nutritional Composition and Heavy Metal Content in Some Edible Mushroom Varieties Collected from Different Areas of Bangladesh. *Asian J. Med. Biol. Res.* **2015,** *1*(3), 495–502. DOI: 10.3329/ajmbr v.1i3.26468.

Mozumder, M. M. H.; Shamsuzzaman, M. M.; Rashed-Un-Nabi, M.; Karim, E. Social-ecological Dynamics of the Small Scale Fisheries in Sundarban Mangrove Forest, Bangladesh. *Aquac. Fish.* **2018,** *3* (1), 38–49.

Mukherjee, N.; Sutherland, W. J.; Dicks, L.; Hugé, J.; Koedam, N.; Dahdouh-Guebas, F. Ecosystem Service Valuations of Mangrove Ecosystems to Inform Decision Making and Future Valuation Exercises. *PloS One* **2014,** *9* (9), e107706.

Neogi, S. B.; Dey, M.; Kabir, S. L.; Masum, S. J. H.; Kopprio, G.; Yamasaki, S.; Lara, R. Sundarban Mangroves: Diversity, Ecosystem Services and Climate Change Impacts. *Asian J. Med. Biol. Res.* **2017,** *2* (4), 488–507.

Queiroz, L. de S.; Rossi, S.; Calvet-Mir, L.; Ruiz-Mallen, I.; Garcia-Betorz, S.; Salva-Prat, J.; Meireles, A. J. de A. Neglected Ecosystem Services: Highlighting the Socio-cultural Perception of Mangroves in Decision-making Processes. *Ecosyst. Serv.* **2017,** *26*, 137–145. https://doi.org/10.1016/j.ecoser.2017.06.013.

Rahman, M. M.; Asaduzzaman, M. Ecology of Sundarban, Bangladesh. *J. Sci. Foundation* **2010,** *8* (1 & 2), 35–47.

Rahman, M. M.; Rahman, M. M.; Islam, K. S. The Causes of Deterioration of Sundarban Mangrove Forest Ecosystem of Bangladesh: Conservation and Sustainable Management Issues. *Aquac. Aquar. Conserv. Legis.* **2010,** *2* (3), 77–90.

Rodrigues, J. G.; Conides, A. J.; Eodriguez, S. R.; Raicevich, S.; Pita, P.; Kleiner, K. M.; ... Villsant, S. Marine and Coastal Cultural Ecosystem Services: Knowledge Gaps and Research Priorities. *One Ecosyst.* **2017,** *2*, e12290. https://doi,org/10.3897/oneeco.2.e12290.

Spalding, M.; Kainuma, M.; Collins, L. *World Atlas of Mangroves*; Earthscan: London, UK, 2010; p. 319.

Sunny, A.R. Impact of Oil Spill in the Bangladesh Sundarbans. *Int. J. Fish. Aquat. Studies* **2017,** *5* (5), 365–368.

Thibodeau, F. R.; Nickerson, N. H. Differential Oxidation of Mangrove Substrate by *Avicenia germinans* and *Rhizophora mangle*. *Am. J. Bot.* **1986,** *73*, 512–516.

Uddin, M. S.; van Steveninck, E. D. R.; Stuip, M.; Shah, M. A. R. Economic Valuation of Provisioning and Cultural Services of a Protected Mangrove Ecosystem: A Case Study on Sundarbans Reserve Forest, Bangladesh. *Ecosyst. Serv.* **2013,** *5*, 88–93.

Walters, B. B.; Ronnback, P.; Kovacs, J. M.; Crona, B.; Hussain, S. A.; Badola, R.; ... Dahdouh-Guebas, F. Ethonobilogy, Socio-economics and Management of Mangrove Forest: A Review. *Aquat. Bot.* **2008,** *89* (2), 220–236.

CHAPTER 5

SUBAH-E-BANARAS AND KASHI VISHWANATH DHAM CORRIDOR: AN ADDITION IN HERITAGE TOURISM AND NEW AGE TOURISM OF VARANASI

PRAVIN S. RANA[1*], SUDHANSHU MANI[2], and SHUSHANT KUMAR KUSHWAHA[2]

[1]*Faculty of Arts, Banaras Hindu University, Varanasi, India*

[2]*Tourism Management, Banaras Hindu University, Varanasi, India*

[*]*Corresponding author. E-mail: psrana@bhu.ac.in*

ABSTRACT

Everything that remains unchanged for a long course of time is believed to be on the verge of culmination. The characteristics of change are rather appreciated in nearly every walk of life. Tourism remains the primary halt to such dynamics. Tourist seeking change must be channelized in some way or other to maintain as well as to boost the tourism influx. Varanasi or Banaras houses the best of its tradition and culture since the inception of these vary two terms.

Heritage and tourism to those heritage sites is very crucial and requires different approach for planning and development. Of course, it can easily be narrated that as a group or separate category heritage tourism does not exist; however, as a strategy for the sustainable development, a yardstick of heritage always be given special place. The city of Varanasi and its heritage have been always attraction for visitors and pilgrims for many hundred years. The existing intangible heritage or tangible tourism products and upgradation in some of existing resources are progressing through HRIDAY projects in Varanasi city, which is altering the regular trends of the city.

As per new age tourists, they display the culture, respect the host destination, and are environmentally conscious. They look to experience and learn rather than stand back and gaze. These new age tourists are participators not spectators also mingling with the locals to make it more bona fide. This initiative was taken by district administration on November 24, 2014 to showcase the cultural epitome of Kashi. The program starts with morning Ganga Aarti with shehnai recitation (classical musical instrumental) and mangalcharana (Sanskrit chanting from Vedas) followed by one hour of yoga session and then two hours of classical music recital by budding and eminent musicians from all parts of Banaras and world. Besides yoga, music ancient Yajna (fire rituals) practice also takes place side by side. The new age tourists and guests generally visit Varanasi for the eternal peace and to feel within the existence of omnipotent, nature, and variety of traditions intermingled together in the life of the city.

The present paper attempts to study the existing and upgradation of two heritage resources, that is, *Subah-e-Banaras* and *Kashi Vishwanath Dham* Corridor that will help to enhance the experience of visitor in respect of pilgrimage tourism, heritage tourism, and new age tourism.

5.1 INTRODUCTION

Mark Twain once wrote, "Benaras is older than history, older than tradition, older even than legend and looks twice as old as all of them put together" (Twain, 1897).

Several sacred and exotic sites situated in its geographical boundary are already well celebrated among tourists and pilgrims. A regular activity of a visitor which includes pilgrims and tourist is accessing the sites like River front Ghats of Banaras, heritage walk in narrow lanes, enjoying street foods, and purchasing local handicrafts. But there are few missing component of originality since long time as reported by visitors.

Taking this missing element, stakeholders including local administration started thinking about creating a new event which can cater international and domestic tourist. This gave birth to the proceeding that is presently called *Subah-e-Banaras* which organized its first show in 2014 at bank of Ganga, at Assighat to exhibit the striking yet traditional way of life in Banaras.

Sri Kashi Vishwanath temple situated at Varanasi is among the 12 other Jyotirlingas. Vishweshwara or Vishwanatha is considered to be the most sacred among all 12 Jyotirlings situated all over India. According to tradition, the merit earned by single visit of this Jyotirlinga is equal to accrued merits by visiting rest of 11 Jyotirlingas across India. The famous *Vishwanathgali* (narrow lane) leads the path to inner temple sanctum through the gate number. It is a narrow lane with shops of local handicrafts, souvenir, Prasad, and local weaved garments displayed on either side. On an average, 3000 pilgrims and tourist visit this place every day and the count rise to more than a million on any special occasion. The movement of pilgrims is often obstructed by stray animal inside the narrow lane as it is very irregular by length and breadth. The corridor plan is a decision to develop the facilities for the pilgrims and visitors keeping their requirement on priority. The 11.6 acres of land has been evacuated to develop the temple and river Ganga face front by bridging the inner sanctum of Vishwanath temple to Manikarnika ghat situated at the Northern part of the temple. Pilgrimage tourism with regards to Kashi Vishwanath corridor expansion is being considered as the major make-over to facilitate the tourists and pilgrims with various newly induced facilities that shall be built inside its premises.

Banaras is bestowed with nearly all the religions of India. Every religion is bestowed with their heritage monuments and sites. It has Hindus heritage temples manifestation and their replicas (12 Jyotirlingas and 12 Swyamb-hulinga), Jain heritage sites (Chhedilal Digambar Jain Mandir, Sanmati Digambar Jain Mandir, and Digambarjain Mandir), Sikh heritage sites (Guru-ka-Bagh Gurudwara and Nirmal Sanskrit Mahavidyalaya), Christian churches as heritage sites (St. Thomas, Red, St. Paul, and David's Church), Muslim heritage sites and monuments (Dhai Neem Kangoore, Ganje Sahada, Chaukhamba, Biwi Razia, Gyanavapi, and Alamgiri Dharahra Masjid), and Buddhist ruins and Stupas (Dhamekha and Dharmrajika Stupa).

New age tourism comprises traveling with prior preparations to the destination which is located away from the daily workplace. Further, it explains the idea of leisure setting which happens to be voluntarily self-indulgent, commercial because there is transaction purchase that is restricted not only to tangible products but also to exotic, novel, sacred, and life transforming experiences.

External environment has a significant role in the evolution of the tourism. A list of many other activities was economically equal to tourism at the initial days but in long run it turned less favorable due to mass tourism factor. In 1950, 1960, and 1970, the global period of mass tourism or old tourism was followed by new age tourism. Banaras is the cultural and spiritual heritage hub for tourists who not only believe just in simple packaged tours but also prefer to dive deep in the "Juice of life" here inside city. This city caters the very basic to premium needs of every visitor according to their demand and hence it has been among the most favored destination for every holiday planner, making it a destination for mass tourism.

Mass tourism, when compared, has more negative impacts than positive, and therefore though it boosts local economy, it also exploits the natural and cultural heritage. The city has maintained and forgoes the effects of it, but this is the high time to discuss the various aspects and proper ways to maintain the balance by practicing sustainability and adding new options in the present scenario.

The best option was to divulge the tourist mass for the better management and proper showcasing of the culture in order to preserve natural heritage of Banaras. Introduction of *Subah-e-Banaras* was considered the very first step toward the many more suggestive to promote and preserve the culture of the city by going back to our roots.

The present paper is attempted to study the existing and upgradation of two heritage resources, that is, *Subah-e-Banaras* and *Kashi Vishwanath Dham Corridor* that will help to enhance experience of visitor in respect of pilgrimage tourism, heritage tourism, and new age tourism.

5.2 LITERATURE REVIEW

The term new age tourism is somewhat connected with the limits of elective tourism. Elective travel is considered to be conventional methodology of Fordist as the dismissal of mass travel, for an increasingly imaginative, specific, and greener Post-Fordist approach considered as practices and estimations of counter-social developments of the 1960s (Conti and Perelli, 2019). New age travelers' ideological duties differ impressively (House, 1994) with this type of the travel industry exercises and encounters that are explicitly characterized by expansive ideological range (Weiler and Hall, 1992). New age tourism is likewise viewed as grasping some type of environmentalism esteems network (Morphy, 1992) additionally create productive "host and visitor" (Smith 1989) associations (Eadington and Smith, 1992) alongside genuine social duty thought (Krippendorf, 1987).

Heritage is a rising entity in the travel industry business giving sightseers' encounters dependent on the substantial and elusive survives from the past. Because of the absence of different potential outcomes, the arrangements found in fringe domains frequently feature the travel industry—especially heritage tourism—as the most significant action to dispatch an increasingly practical advancement process (Fonseca and Ramos, 2012). The issues of heritage tourism in Varanasi and its method of reasoning of the engraving as a world legacy site are considered by Singh and Rana (1993, 2017, 2019) portraying how the city of Banaras (Varanasi) is significant from legacy just as a hallowed scene.

5.3 THE MODERN/NEW TOURIST

The tourist today is altogether different from previous patterns. The best organizations in the travel business are those that tackle the situation utilizing technology, inventive promoting programs, better preparing of staff, and by creating closeness and comprehension of its clients/visitors. The contrast in tourism designs in the following century will be progressively identified with what travelers are looking for in a tour experience than in the way they travel. The present explorer, the well-behaved or unbridled hiker, is typically educated, instructed, and in general, completely assured of what the person is going to extract from their tour. The new age tourist is more comfortable by using mobile app rather taking help from travel agents (Table 5.1). For them, the streamlining of time and cash is the key and they get ready for their excursion by investigating the place through the web and the encounters of companions and individual voyagers. Broadcasting media acted as a catalyst to this move and further energized the quest for encounters through the advancement of an immense scope of the way of life/experience programs that have developed more as of late into experiential touristic docu-dramatizations alluded to as reality developers.

The focus is being redirected in investigating new outskirts where conventional ideas did not permit. In Tourism, Technology, and Competitive Strategies, Poon (1993) talks about the adjustments in buyer conduct and qualities that are the basic main thrusts for the new travel industry.

The new traveler is experienced, increasingly adaptable, autonomous, quality cognizant, and harder to please. They are more often interested and generally focused for ecological balancing activity, deep concerned in displaying all respect in the tangible and intangible heritage of host nation; overall, they can be termed as the submissive and indulgent lot rather gazers. Things that could never show up on the rundown of the "mass" vacationer, for example, experience, getting off the beaten track, and blending with local people are currently the establishments of the new visitor encounters. Ordinarily, these voyagers are getting some distance from movement and lean toward an abnormal state of contribution in the association of their outing.

TABLE 5.1 Comparison of Old and New Tourist.

Old tourist	New tourist
Stereotype	Experience something exotic
Mass tourist	Self-propelled
Nomads	See, enjoy, and preserve
Been there, done that	Enjoy the core of it
Encounter	Existence
Superiority	Appreciation
Admire attractions	Admire sports
Preventive	Happening
Eat in hotel	Enjoy local fare
Congruent	Compound
Use travel agents	Use Mobile app
Information about destination by print	Information by review web sites

Source: Based on Poon (1993) Tourism, Technology and Competitive Strategies.

Travel is never again a curiosity to the new vacationer. Studies bolster what industry officials have been seeing throughout a previous couple of years. Individuals expect more out of their excursions than they used to, and they are progressively gutsy. Overviews did by the Canadian Tourism Research Institute demonstrate a high level of enthusiasm for escape excursions, ecotourism, cultural tourism, and consolidating a professional trip with a pleasure trip. Throughout the following 10 years, the travel industry items and attractions should take into account guests who are all more requesting and separating, just as increasingly dynamic and progressively intentional in their decision of destination. There will be a move in accentuation from latent amusing to dynamic learning, and the quality and validity of guest encounters will be vital to future accomplishment in a focused market.

An Acronym that is important to depict the "new" traveler is REAL, which represents Rewarding Enriching Adventuresome Learning Experience. A key supporting idea for REAL tourism is the genuineness of experience, which is regularly identified with the habitat and culture and seen to be unaffected by "mass" travel. New tourists like to be viewed as voyagers and not vacationers. Some particular focuses that should be remembered while managing the new tourist are: This kind of voyager requires a totally extraordinary showcasing approach. They keep away from regular polished showcasing instruments and want to utilize solid sources, for example, informal referrals, their very own free research and confided in distributions; they want encounters rather than items and administrations; they can be called experiential voyagers, they reach out over all age gatherings and conventional market sections.

5.4 OBJECTIVE

1. To identify the active role of the *Subah-e-Banaras* as an event to facilitate new age tourism.
2. To forecast the impetus of the pilgrimage and heritage tourism post *Kashi Vishwanath corridor* construction.

5.5 METHODOLOGY

Primary and Secondary data have been collected because of their general nature in the field of tourism research. Participatory observation was used in order to collect the primary data whereas the secondary data were availed by the government and various research organizations; the local administration and the *Subah-e-Banaras* committee were a great help in the collection of secondary data. Participatory observation conducted was rigorous and continuous in nature as the few of the visitors were unknown to both the language that is Hindi and English. In order to observe the instant reaction of the tourist, we were present throughout the event and interview of the repeated respondents that were also recorded as they were unable to fill the questionnaires at that very point of time. In process to check the biasness of the observations, different age groups of tourists were interviewed. During the interview, questions were asked ranging from the experience of the event to the after effect of it; motivation to visit this event again and the whole holistic views about this city after the addition of this event in the to do list of the visitors. The same observation and field study were conducted to know present status of Vishwanath Temple corridor, its future plan, and some observation and interview with local and pilgrims about the future development.

5.6 HERITAGE

The term has broader meaning to communicate and it is totally perception based on so many criteria. To summarize in more layman term, it can be simply defined as everything which was entrusted by our ancestors or inherited from them may include under this category. It could be structures, landscapes, tradition, objects, etc (Singh and Rana, 2017).

These all were termed combined which was later distributed in two different segments with their own criteria intact which was framed by World Heritage Council (WHC) a special body instituted under UNESCO in 1972 convention with the main aim to protect the world cultural and natural heritage.

5.6.1 CULTURAL HERITAGE CRITERIA

Article 1 of the convention states further the following items come under this criterion:

1. Architectural monuments work, works of monumental sculpture and painting, elements, or structures of an archaeological nature, inscriptions, cave dwellings, and combinations of features having outstanding universal value from the point of view of history, art, or science.
2. Groups of buildings separate or connected because of their architecture, homogeneity, or landscapes.

3. Sites that are works of man or combined with nature and of man with universal value from the historical, aesthetic, ethnological, and anthropological point of view.

5.6.2 NATURAL HERITAGE CRITERIA

As per Article 2 of the convention goes under this model:

I. Natural highlights comprising of physical and natural arrangements or gathering of such developments, which are of exceptional widespread incentive from the stylish or logical perspective;
II. Geological and physiographical developments and definitely depicted regions which comprise the natural surroundings of compromised types of creatures and plants of remarkable widespread incentive from the perspective of science or protection;
III. Natural destinations or exactly outlined characteristic zones of exceptional widespread value from the perspective of science, preservation, or normal magnificence.

5.6.3 RATIONALES FOR INCLUDING VARANASI AS HERITAGE CITY IN THE UNESCO WORLD HERITAGE LIST

5.6.3.1 CONSTITUTE A UNIQUE ARTISTIC AND AESTHETIC ACCOMPLISHMENT

The city represents a unique natural shape with the Ganga river flowing northerly for 7 km of its course that shapes crescent and city has grown on left bank in circular form around it. The natural ecosystem is preserved by the area on its right side which serves as a flood plain. Thus, the combination of its either sides represents the cultural and natural heritage and that is also unique in whole of India.

5.6.3.2 RECORDING CONSIDERABLE INFLUENCE AT A CERTAIN PERIOD

Varanasi as a history has special role in promoting education, debates, and dialectics both religious and spiritual and manifestation of holy centers of pan-India sites. Sanskrit and Ayurveda are believed to be here since fifth century BCE. All major holy centers of India were believed to be replicated here before 12th century. It can be undoubtedly termed as mini-India and also perceived as the cultural capital of India.

5.6.3.3 EVIDENCE OF A DISAPPEARED CIVILIZATION

Sarnath where Buddha gave his first sermon, "Turning the wheel of law" c.a.528 BCE and later during third century BCE when king Ashoka built a monastery township there that was there till 12th century CE and then destroyed. The second was the excavation of archaeological findings from the Rajghat area of Varanasi that was further supported by Robert C. Eidt (1977) on the basis of scientific analysis that they were concluded to be belonged to 800 BCE to CE 800. It was derived out of from the results that residential settlement prevailed at this place at those time spans was continuous. The small evidence of farming and domestication of cattle was a sign of pastoral economy and the testimony to oldest occupied modern cities in the world.

5.6.3.4 ILLUSTRATE A SIGNIFICANT HISTORICAL PERIOD

From sixth century CE to 12th century, the city had established itself as the most popular holy center for Hindus. The images of various deities were established during this period, which now reaches over 3000 Hindu shrines and a few Buddhist, Jain, and Sikh shrines. Muslims shrines prevailed later but now they have reached over 1300 that is more than any Islamic site.

5.6.3.5 CONSTITUTE AN OUTSTANDING EXAMPLE OF A TRADITIONAL WAY OF LIFE

This is a city that celebrates life each and every day of it. It is considered the veritable jungle of fairs and festivals with respect to distinction, sacred sites, time, variety, overseers, and side-shows. The Ganga being the center of it has always been associated with the sacred bath in order to have a fruitful death and attain transmigration. Presence of "dying homes" charitable homes, 84 *ghats* along the Ganga are equally contributing to its characteristics. The silk weaving, sari making, wood and terracotta handicrafts, and toy making comprise the continuity of historic-cultural tradition. The *Banarasi* lifestyle is the blend of the entire above-mentioned ingredient and holds a great pride among the natives. The lifestyle has manifested itself in musical tradition known Banaras Gharana (style). The soil is the home to many renowned artists and their performance has been the testimony to this fact since ages.

5.7 NEW AGE TOURISM AND *SUBAH-E-BANARAS* ROLE AS AN ADD ON TO CULTURAL HERITAGE

Assighat which is located southern among ghats, the confluence of Ganga and Assi river tributary, is considered to be sacred for holy bath during full moon. Numerous mythological saying has been associated with this place and the ghat remains among the top celebrated area among the tourists. The river bed has been shifted away from the ghat in due course of time and deposition of flood plain has taken its place making it greener than ever. The steps of 19th century Sangameshwara temple have always been a spot for local tourists to sit and spend their time. The students of Asia's largest residential Banaras Hindu University can be spotted spreading the messages among the locals and the tourists for different reasons irrespective of their personal benefits which makes it again a spot of great importance socially.

The entire residential area of Assi is acquired by tourists who not only come Banaras to get a glimpse of it but also stay here and learn the different arts expanding from Sanskrit to music and dance to devotion. *Subah-e-Banaras* is not adding anything extra what Banaras has to offer for the seekers of light and moksha rather it is the platform which delivers the entire taste of ancient and modern Indian culture evolved during different phase of time in Banaras but were later discontinued due to mixing and acquiring of different culture and mindset to established way of life. The dilution of culture and tradition of this place is being concentrated by this vary organizing. The involvement and experience of this event with tourist and localities is what make this a very suitable example for new age tourism. This is the platform that not only entertains you rather it takes your breath away by giving life changing experiences by showcasing the pragmatic aura of life.

5.7.1 WHY SUBAH-E-BANARAS IS TOP OF TO DO LIST?

Ganga Aarti (oil lamp ritual): Ganga is the life line of this eternal city. The river flows from north to south direction in this city and water of the Ganga is considered as purest spiritual energy equivalent to elixir. The *Aarti* on the banks of Ganga with the sun rising in the east remains the first event that marks the beginning of new day which brings life to everyone on the earth (Fig. 5.1).

FIGURE 5.1 *Subah-e-Banaras* activities—Arati.
Source: Picture taken by Author.

Yoga: After thanksgiving to Ganga, the next step which follows is the self-building as it has been rightly said that "a sound mind lives in a sound body". A daily one-hour yoga session is also conducted to impart authentic knowledge of this very lifesaving mantra to the tourists.

Yagna (vedic fire ritual): Fire ritual is the practice which is continued in Indian societies since ancient times and considered as very close to nature. Since starting of the life, this practice has been there in order to achieve some natural and spiritual gain in daily life. The concept behind this is it works as the bridge between the living and the almighty (Fig. 5.2).

Cultural events: Sustainability and the society participation are very important prospects for any event to make it successful. The local artist of Varanasi is invited on daily basis to perform in front of all the tourist and dignitaries for two important reasons:

1. To showcase the rich and varied heritage of this city and also the relative evolution in the field of performing art and culture.
2. It also works as the marketing platform for them to meet and understand new business ideas and proposals for the future assignments.

FIGURE 5.2 *Subah-e-Banaras* activities—Yoga, fire ritual, chanting of scared verses.
Source: Picture taken by Author.

To sum it up the entire event is a great support to strengthen the cultural stagnation of Banaras. The tag of repeated and stereotypic experience is highly dangerous, and this is the high time to come up with the more of these ideas in order stay on the top list of cultural destinations because repetitions are always treated as spoiler to the authentic experiences. However, the exact quantum of the impact caused on the tourist influx is not very clear yet as there are plenty of places which shadow the importance of *Subah-e-Banaras* in general. A study of decade can help to understand the same more precisely and exact.

5.7.2 KASHI VISHWANATH DHAM CORRIDOR

5.7.2.1 THE OVERVIEW

Development is in a way considered a double-edged sword, while the primary aim of the one edge remains the same as of similar to any normal sword except sometimes when the second edge can be fatal to one who holds it. Since the ages *Ganges* and the *Vishwanath* (Shiva in form of lord of universe) are synonyms of *Kashi* and this ideology remains the prime reason of urban agglomeration which was expanded and synchronized in the vicinity of these. *Ganga the life giver* and *Shiva the destroyer or maintainer* have their cult among Indians (mostly) Hindus, and this combination of the pilgrimage and spiritual journey remains the priority of the devotees.

Pilgrimage tourism remained on its regular track despite of the various small blockages present in the route to inner sanctum of the *Vishveshwara* temple. When we consider the entire compound of the *Kashi Vishwanath* being among the oldest of its kind which is not just an ordinary compound rather it is the conglomeration of private residences, bunker shops, stray animals, and a mosque which was built in late 17th century (Fig. 5.3); all of them have their unique importance which is testified by the increasing number of pilgrims year after year. "*Banaras* has been known for its encroached *galis*" which is considered significant as the major pull factor of the increasing footfalls.

5.7.2.2 THE PLAN

The Vishwanath Mandir corridor project costing ₹6 billion was initiated on 11.6 acres land which is equivalent to 47,000 m^2. *Vishwanath Dham* comprises main temple at the western end of which 350 m is the *Ganga* barricaded by three well-known *ghats* named Manikarnika, Jalasen, and Lalita (Fig. 5.4). This corridor pathway cleaning made 42 temples appear which were earlier disappeared due to encroachment and the illegal construction surrounding main temple since centuries, nearly 250 private properties were purchased and demolished by the executive committee which were housed either by individual or families (skparishad.org, 2019). The proposed plan of this corridor defines the entire area with magnanimous grandeur and elite feeling for the pilgrims in the time to come. A large open space is fronting the temple or *Mandir Chowk* with a small passing space for *Gyan Vapi Mosque* which is adjacent to temple. As per news published in Hindustan (Hindi Dainik, dated June 8, 2019), the construction of the Vishwanath temple will be completed in four phases.

FIGURE 5.3 View of Vishwanth temple corridor (under construction).
Source: Picture taken by Author.

The first phase will expand the boundaries around the Vishwanath temple; the temple will be ready around. In the second phase, there will be development works in the gates of the temple between second and third lane to about 10,000 m^2. It includes the restoration of mythological and ancient temples. A vast gate will be made. Facilities will be developed for the devotees in the temple complex and the premises. In the third phase, the corridor will be built from Lalita *Ghat* on the Ganga coast to Manikarnika *Ghat*. Accordingly, the pumping station of Jalasen *Ghat* will be underground. At the same time, a huge platform will be formed over the Manikarnika *Ghat* which will connect the temple directly. In the last phase, work will be completed between Lalita *Ghat* and Nilkantha Gate. Development of 42 ancient temples hidden inside the buildings will be developed based on their historical richness and grandeur. A board displaying the wonderful architectural and craftsmanship will be arranged there inside the premises of these temples. The new compound shall remain the ground zero for optimum facilities required by pilgrims. These contain guest houses, Yatra Seva Kendra's, *Yagnsalas* for *yagna* and *havan*, hospice, museums, and library also with special attention to adequate number of public toilets. Escalator facility shall be made available at Nepali (Lalita) *Ghat* with an additional pathway from Manikarnika *Ghat* to Temple has been proposed for the elderly and differently abled pilgrims.

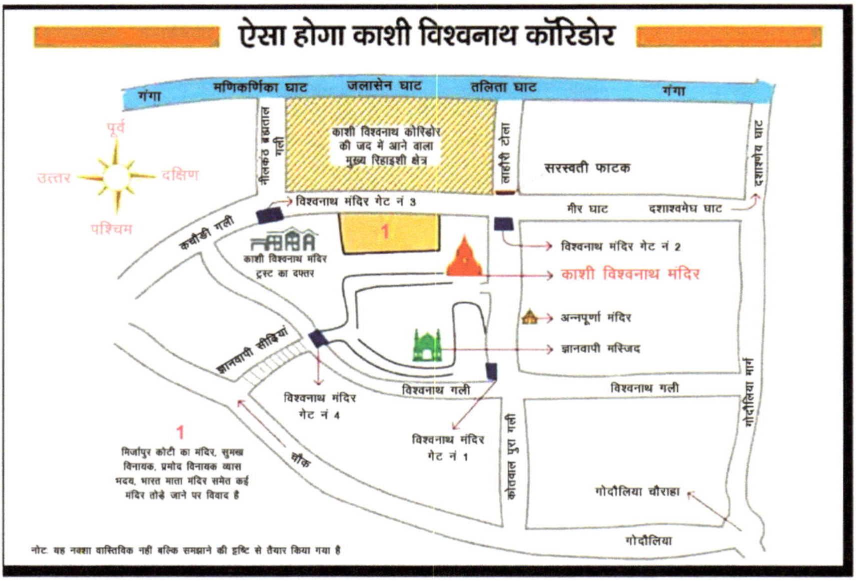

FIGURE 5.4 Proposed map and design of Kashi Vishwanath corridor.
Source: www.thehindu.com/society/varanasi-by-design-vishwanath-dham

The facilities projected will boost the influx of tourist and to maintain that the extensive security arrangements. The quad level security service shall work along the Integrated Command Control Center with a separate building dedicated to them. First level shall witness use of Telemeter which will fire an alarm in case of emergency followed by Hydraulic based Bollard System which has a significant role in security of such public places. Third level will witness Access Control System which is responsible for x-rays and scanning of objects/humans entering the temple premises along with Air Surveillance System at level four to crack down any irregularities, if found within 200 m of air space of the Vishwanath Dham Corridor. Also, a wider emergency gate shall be constructed which will give access to ambulance directly to the temple premises in case of emergency.

5.7.2.3 FUTURE REVERBERATIONS AND IMPACT ON PILGRIMAGE AND HERITAGE TOURISM

The cultural heritage is considered priceless and irreplaceable and this remains constant for every nation. Any loss through disappearance and deterioration even slightest of them is destitution of the heritage of the entire mankind. If an estimated rough data is to be believed than present arrangement of the *darshan* (meeting the deity) is out-turned by 3000 pilgrims daily and the number crosses the count of 100,000 on certain auspicious occasions. With the inauguration of the corridor, the wider inner area will completely eradicate the lumpy encroachment issues, hence making the *darshan* fruitful yet comfortable together.

5.8 EPILOGUE

The perception of behavior of visitor to this eternal city of Varanasi is changing. Earlier only expectation from the city was some cultural and religious sites with lifestyle of people but now as number of repeated visitors are increasing (Table 5.2), Varanasi was facing shortage of new attractions to be offered. After introduction of *Subah-e-Banaras* and coming project of Kashi Vishwanath Temple dham corridor is already started attracting so many varieties of pilgrims, visitors, and tourist in the city. Being parliamentary constituency prime minister of India, it also attracts visitors, investors, and planners to study and offer new age tourism for all.

TABLE 5.2 Varanasi: Tourist Influx (2013–2017) in million

Year	Domestic/International	Uttar Pradesh	Varanasi
2014	Indian	23.08	5.2
	Foreign	1.88	0.287
2015	Indian	22.32	5.41
	Foreign	1.82	0.302
2016	Indian	22.23	5.6
	Foreign	1.85	0.312
2017	Indian	26.12	5.95
	Foreign	2.28	0.335
2018	Indian 26.7 6.2		
	Foreign	2.37	0.416

Source: UP Tourism, 2019.

Of all the encounters and interviews that were conducted ethnographically in a systematic and channelized way pointed out some obvious points and inducted some significant research gaps indeed.

Arrivals on an average are inclining year by year and the stagnation of cultural prospect cannot be proved there of whereas the specific motivation of visiting Banaras only for the participation in *Subah-e-Banaras* was also not proven as absolute. Interviews on the Day 1; 2 September, 2018 morning 6 am/Varanasi- Assi Ghat "Great feelings. "Want to visit again and again" exclaimed a domestic tourist coming all the way from Kolkata just to witness the grandeur of this early morning festivity. Interview on the day 10; September 12, 2018 morning 7 am/Varanasi-Assi Ghat. "We came to witness the spiritual essence of Varanasi, thanks for the *Subah-e-Banaras* as it was icing on the cake for us, good luck and great wishes for the entire team of this event" were the remarks of Cindy Morkel who came all the way from United States to visit India and in that Banaras specifically.

Local people and new age visitors are eagerly waiting for completion of the Vishwanath Dham corridor to have completely different experience of spirituality, religion with development.

KEYWORDS

- **new age tourism**
- ***Subah-e-Banaras***
- **Kashi Vishwanath Corridor**
- **heritage tourism**
- **ghats of Banaras**

REFERENCES

Bibhudatta, B. Design Resource Culture, City and Crafts of Varanasi. http://www.dsource.in/resource/culture-city-and-crafts-varanasi/credits.

Conti, G.; Perelli, C. Traditional Mass Tourism Destinations: The Decline of Fordist Tourism Facing the Rise of Vocational Diversification. Governance and Sustainability in New Tourism Trends, 2019. http://media.planum.bedita.net/49/58/Conti_Perelli.pdf.

Eadington, W. R.; Smith, V. L. Introduction: The Emergence of Alternative Forms of Tourism. In *Tourism Alternatives*; Smith, V. L., Eadington, W. R., Eds.; University of Pennsylvania Press: Philadelphia, PA, 1992; pp. 1–12.

Fonseca, F. P.; Ramos, R. A. Heritage Tourism in Peripheral Areas: Development Strategies and Constraints. *Tourism Geographies* **2012,** *14* (3), 467–493.

House J. An Alternative Form of Alternative Tourism? Unpublished Thesis, 1994. http://www.arasite.org/pspage2.htm (accessed April 12, 2019).

Krippendorf, J. *The Holiday Makers. Understanding the Impact of Leisure and Travel*; Butterworth Heinemann: Oxford, 1987.

Morphy, H. 'The Art of Basketry: Aesthetics in a Cross-Cultural Perspective'. In *Mowat, Morphy, Dransart*, Eds.; 1992; pp. 160–168.

Poon, A. *Tourism, Technology and Competitive Strategies*; CAB Wallingford: Oxford, 1993.

Rana, P. S.; Singh, R. P. B. Behavioural Perspective of Pilgrims and Tourists in Banaras. In *The Tourist—A Psychological Perspective*; Raj, A., Ed.; Kanishka Publ.: New Delhi, 2004; pp. 187–206.

Singh, R. P. B. 1993/2018. The GaṅgāGhāṭs, Varanasi (Kashi): The Riverfront Landscapes. In *Banāras (Vārāṇasī): Cosmic Order, Sacred City, Hindu Traditions*; Singh, Rana P. B., Ed.; Tara Book Agency: Varanasi; pp. 65–102. Rev. pp. 1–77.

Singh, R. P. B.; Rana, P. S. Varanasi: Heritage Zones and Its Designation in UNESCO's World Heritage Properties. *Kashi J. Soc. Sci.* **2017,** *7* (1–2), June–December, 201–218.

Singh, R. P. B.; Rana, P. S. Faith and Place: Hindu Sacred Landscapes of India. In *The Routledge Handbook of Place*; Edensor, T., Kothari, U., Kalandides, A., Eds.; Routledge: London, 2019a; ca. pp. 141–154.

Singh, R. P. B.; Rana, Pravin S. Visioning Cultural Heritage and Planning: Banaras, India. In *Routledge Companion of Global Heritage Conservation*; Vinayak, B., Sandmeier, T., Eds.; Routledge: Abingdon, Oxen, UK, 2019b; pp. 175–200.

Singh, R. P. B.; Rana, P. S.; Kumar, S. Intangible Dimensions of Urban Heritage: Learning from Holy Cities of India. In *The Routledge Handbook on Historic Urban Landscapes of the Asia-Pacific*; Silva, K. D., Ed.; London; pp. 65–84.

Srivatsan, A. (2019). Varanasi by design: Vishwanath Dham and the Politics of Change-The Hindu, accsseed on 28.7.2020 tps://www.thehindu.com/ society/varanasi-by-design-vishwanath-dham- and-the-politics-of-change/ article26607193.ece.

Twain, M. Following the Equator, Chapter 50-On the Road to Benares, 1897. http://www.online-literature.com/twain/following-the-equator/0/.

UP Tourism Statistics Report. 2014. http://uptourism.gov.in/site/writereaddata/Monument-small-chart-2014-to-2018.pdf.

van Hove, N. Globalisation of Tourism Demand: The Underlying Factors and the Impact on Marketing Strategy. In Keller, P., Ed.; *Globalisation in Tourism, Tourist Review* **1996,** *4*, 6–7.

Weiler, B.; Hall, C. M. *Special Interest Tourism*; Belhaven Press: London, 1992.

Hindustan. (Hindi Newspaper); Vishwanath Mandir Corridor: Varanasi, June 8, 2019.

CHAPTER 6

HERITAGE AND SUSTAINABLE DEVELOPMENT: CAPACITY BUILDING THROUGH TOURISM

ISABEL VAZ DE FREITAS[1*] and MICHAŁ RYSZARD KOSKOWSKI[2]

[1]*Universidade Portucalense, Porto, Portugal, Research Group on Economics, Management and Information Technologies—REMIT*

[2]*Szkoła Główna Handlowa w Warszawie (Warsaw School of Economics), Warszawa, Poland*

[*]*Corresponding author. E-mail: ifc@upt.pt*

ABSTRACT

Heritage protection becomes an ever-greater challenge due to constant and numerous threats by natural elements and the sole passage of time. This deterioration is increased by other stress circumstances such as economic, social and development pressures, climate change, war, social and religious conflicts, and a number of other factors related to heritage use.

Heritage and historical landscapes should be preserved not only because of their role as an important testimony to bygone times and the source of a spirit of place, but because heritage and culture are important factors of territorial and economic development. Moreover, tourism is one of the sectors of economy that could improve the conservation of heritage and maintenance of cultural diversity. The study will facilitate an author review to complete the picture of the current situation and to find gaps; a documental analysis in order to capture the main concepts, compile the risks and opportunities from selected international and European documents as well as interviews with selected heritage stakeholders and professionals in relevant fields.

6.1 INTRODUCTION

Settled landscapes around the world are rich in resilient cultural elements (tangible and intangible), a sum of activities of earlier human generations. This continuous process results in an accumulation of cultural layers are expressions of a social diachronic perspective that mark the inheritance of today's living societies.

These layers are testimonies of emotions, beliefs, traditions, behaviors, and needs created and collected through human action with the passage of time. In this context, heritage as a cultural resource, material, and immaterial, has been among the key elements of most aspects of urban and rural landscapes. Foremost, heritage includes, as detailed in the Québec declaration (UNESCO, 2008), the tangible and intangible heritage, elements that carry memory. These tangible and intangible elements bequeath places with identity and character.

History, heritage, and culture are the past, present, and future landscape builders, working in a close interaction with communities and the nature itself. Monuments, sites, and intangible heritage are inseparable from the history and their setting. It is the time and places that give a character to heritage, and increase its intrinsic values (UNESCO, 1964).

It is necessary that cultural heritage, tangible or intangible, in macro and in micro places, should be protected and conserved for the use of future generations. This simple rule ensures that the knowledge contained in heritage and its capacity to testify to the diversity and history of the planet Earth is preserved. For the present generations, the cultural heritage is a human right in its diversity, access, and use (Universal Declaration of Human Rights, 1948).

As a testimony of human history and as a human right, the cultural value of heritage goes beyond its value as an economic asset. It is not only intrinsic but also extrinsic. As Fredheim and Khalafor, among other authors, mention that experts have different measures for the heritage values. Economists recognize its dynamic use and the cultural professionals distinguish the heritage values when heritage is both in use (use value) and when it is not in use (the non-use value). In that case, the value is reflected in the significance concept and the criteria (ICOMOS, 2013).

It is possible to find a balance between the economic and the cultural and historical perspective and contribute to sustainability by connecting the cultural and heritage conservation management practices with the economic exploitation of heritage. Heritage is an important component in economic development, in particular through tourism and other commercial activities, notably in rural areas, depopulated places, or economically and socially deprived zones. In this context, it is a priority to highlight the role of cultural conservation in economic and territorial development and to identify adequate methodologies to implement the economic perspective within the cultural management and preventive conservation practices to mitigate against heritage damage or destruction. Otherwise, the damage or disappearance of the cultural or natural heritage is a question of impoverishment (UNESCO, 1972). Moreover, the economic development of heritage as a territorial resource could be a very stimulating tool for its protection and conservation.

In this chapter, capacity building is understood as the capacity to promote abilities and competences development to achieve concrete objectives (ICOMOS, 2013). The main objective of the text is to develop sustainability and heritage conservation through a coordinated program between stakeholders.

As a research work integrated in REMIT (Research on Economics, Management and Information Technologies) at Portucalense University, it is dedicated to the study of sustainable heritage tourism and to debate methodologies that might improve the process of building stakeholder capacity to improve, in a formal or informal way, the sustainability of heritage tourism. The project intends to develop the following objectives:

a) Study and debate international charts and conventions, and the state of the art knowledge in order to create a conceptual framework based on the solid understanding of the concepts: "authenticity," "preservation for future generations," "heritage safeguarding," "universal values of cultural heritage," "heritage significance," "integrity," "identity," "diversity," "sustainability" of heritage use, "risks," "tourism pressure", and "tourism globalization."
b) Identify risks arising from mass tourism, globalization pressures, and different uses of heritage via press and literature review and a close analysis of selected case studies.
c) Identify heritage stakeholders who are also involved with tourism development (such as destination management organizations in historic cities, nonprofit organizations that manage heritage sites, administrators of cultural sites, and monuments that are open to

the public as tourism attractions) and who could benefit from and contribute to the process of capacity building.

After this first step, in the near future we will be able to conclude and explore the perspectives and interests of those different stakeholders, their objectives, and strategies through semistructured face-to-face interviews or, where applicable, open-ended online questionnaires, in order to understand how tourism could improve the preservation of heritage, respect the diversity, and authenticity of the diversity of cultural values. Primarily, therefore, this project will facilitate an author review to complete the picture of the current situation and to find gaps and areas in need of theoretical support. A documental analysis will follow in order to capture the main concepts, compile the risks and opportunities of risk mitigation from selected international and European documents as well as stakeholders and professionals in relevant field selection. The documental analysis will identify the keywords that indicate the heritage values that could be disseminated in the proposed way to contribute to its safeguarding, conservation, and protection while improving the sustainability of its use.

Specific international and European documents, charts and declarations will help to deliver the above objective, such as Resolutions of the International Symposium on the Conservation of Smaller Historic Towns, at the 4th ICOMOS General Assembly (1975), Charter for the Conservation of Historic Towns and Urban Areas (1987), The International Cultural Tourism Charter (1999), the Nara Declaration (1997), ICOMOS (2013), International Cultural Tourism Charter—Managing Tourism at Places of Heritage Significance (1999), the Xi'an Declaration (2005), Xi'an Declaration on the Conservation of the Setting of Heritage Structures, Sites and Areas (2005), Faro Convention (2005), Council of Europe (2005), the European Charter for Sustainable Tourism in Protected Areas (2007), UNESCO (2008), the Charter for Interpretation and Presentation of Cultural Heritage Sites (2008), The Charter on Cultural Routes (2008), the Paris Declaration on Heritage as a Driver of Development (2011), ICOMOS (2011), and The Paris Declaration On Heritage as a Driver of Development (2011).

6.2 HERITAGE VALUES AND RISKS

All heritage tangible and intangible adds to the essence of places, landscapes, and memories, and represents aesthetic, historic, scientific, social, or spiritual values that justify its protection and preservation. The elements that give significance, importance, and emotion to places are transmitted from generation to generation, constantly recreated by communities and individuals as a response to what happens in the world around them and to the passage of history.

Those territorial elements, material or immaterial, are carriers of memories that are easily forgotten, destroyed or degraded. In this context, the debate must include the heritage conservation and valorization and the use by economic activities, in particular, tourism. Tourism and the other economic activities as well as the new uses of historical sites and monuments could be a threat but they could also improve the conservation and preservation of heritage (tangible or intangible) and promote the maintenance of cultural diversity.

Heritage is authentic, testimony of identity, significant, diverse, and coherent but it is also fragile, nonrenewable and irreplaceable, and often exposed to various threats. It is in these threats that the challenges of policymakers are crucial. Alteration, deterioration, disappearance, destruction, disruption, degradation, damaging, neglecting, transformation, and disrespect are keywords related with the heritage fragility emphasized in the successive international declarations and conventions (Fig. 6.1).

Natural causes are an enormous danger to heritage but the risks caused by the action of man are even more unpredictable. The natural reasons that can affect heritage include climate change, natural disasters, and the natural passing of time. They are exacerbated by other circumstances such as economic, social pressures, armed conflict, and cultural reasons.

Using the documental analysis of international charts and declarations, it is possible to identify other stress conditions that can be categorized into socio-economic and cultural. Inside the socio-economic category, mass tourism, globalization, economic pressure, speculation, social transformation, urban development, land use, pollution, traffic, and displacement of communities and abandoned places are the most-cited risks in international documents (Fig. 6.1). Among cultural factors it is not just the compatible new uses of heritage, its removal or demolition, loss of identity and character, disappearance of cultural practices, and loss or substitution of uses and values that are the most commonly cited threats in international documents (Fig. 6.1).

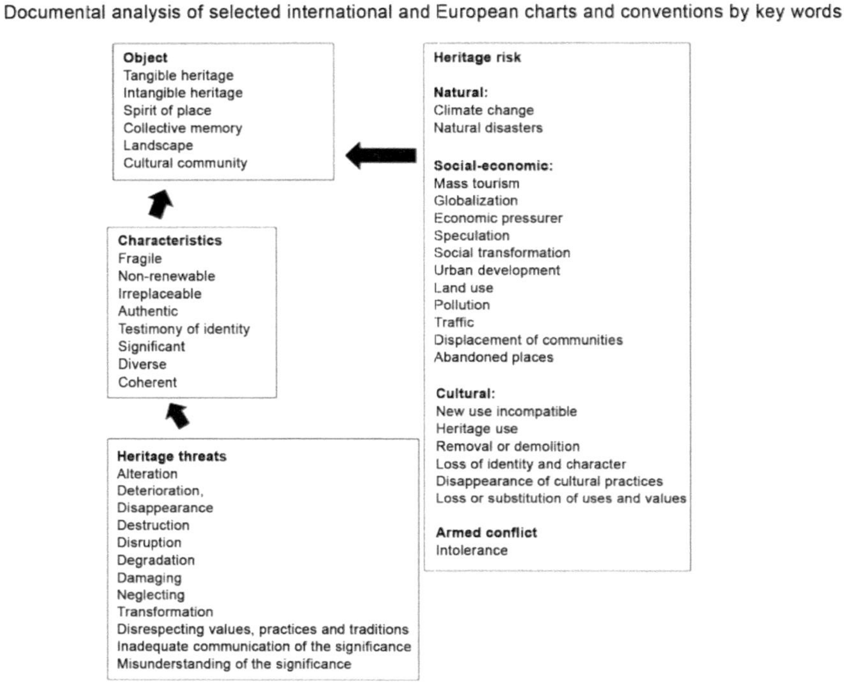

FIGURE 6.1 Heritage and the main risks.
Source: Diagram prepared by authors.

Necessary measures must be taken to ensure the safeguarding, protection, conservation, preservation, and prevention of negative impacts on the tangible and intangible heritage of the landscape, under the consequence of its loss, as soon as the economic evaluation of the territories has taken place.

Losing or decreasing the value of the place or the uniqueness of its character, its authenticity, the diversity of its attributes, and its deep significance, all those losses could jeopardize the meaning of a visit. It would make sense to issue a warning that losing or damaging heritage, one of the most important resources in the local and regional economy could negatively affect the motivation of local people to stay and in consequence lead to their abandoning the given territories. Without the stimulated involvement of local communities, places are affected by lethargy, inactivity, and forgetfulness of memory and heritage. The resulting local low self-esteem must be counteracted.

Heritage protection becomes an ever-greater mission due to constant and numerous threats. Policymakers have to focus their action on the particularity of what identifies their territory and what gives it value by maintaining the spirit of hospitality that has given so much value to tourism, but at the same time they should implement measures to promote sustainability. In addition, special care and work should be undertaken to guarantee such heritage values as authenticity and identity. To explain the importance of authenticity, this concept refers to value-added travel, unprecedented experiences, discovering authentic places, discovery of traditional lifestyle, the desire to make the lived experience a way of life, identifying a population group, discovering vernacular architecture, and discovering of some unknown traditions.

Another risk which heritage may have to oppose are its new and incompatible uses and practices, loss or substitution of uses and values, and the removal or demolition of historical buildings, as well as the increase of new constructions. This loss of uses or values (tangibles or intangibles) could accelerate the movement of communities and the consequent loss of identity. Again, depopulation could have a serious impact on the territory.

In the interior of the country and in many rural areas, heritage often suffers neglect due to the vanishing of cultural practices and abandonment in the face of the departure of local communities, sometimes in relation to the processes of gentrification. Local decision makers need to be aware of the risks of gentrification and implement specific control systems to counteract it (ICOMOS, 2011).

It is necessary to limit or prevent the departure of local people, to promote their establishments, and to attract younger generations. Otherwise, the population decline and the changing social and economic conditions increase the risk of heritage damage or the destruction caused by the lack of economic investment to support its conservation or restoration.

Another factor that causes local communities to disperse is the incapacity of regional and local economies to create opportunities for entrepreneurship and to combat poverty. Policymakers actually are confronted with real territorial problems resulting from the fast economic development, in association with reduced planning for the sustainable development of places and insufficient measures to protect heritage and culture. It is very important to notice that the destruction of heritage, the identity and authenticity lost are correlated with economic, cultural, social, political activities and they threaten the wealth of the territory, as well as its values and significance. Contrariwise, heritage and culture could be treated as assets for local and regional development.

The risks caused by human actions, in particular, the development of mass tourism and the problems resulting from the fast urban development that lead to economic pressure and speculation are the most cited factors in the international documents. The increasing number of tourists generates a massive heritage use, in particular at historic urban centers, at World Heritage sites, and monuments. The increased tourist traffic affects urban development, worsens traffic conditions, and creates extra pollution.

Tourism opened a continuous process of global communication that carries social and cultural transformations and new possibilities of acculturation, creativity development, and cultural fusions. It is the passage of times that occurs as a living being. The new and the traditional may live together because together they enhance economic development and create cultural and social possibilities. But the threat of deterioration, disappearance, and destruction of tangible or intangible heritage is very present in this global process (UNESCO, 2003). Globalization, without concern for planning, vision, strategic thinking, and openness to the global world present itself as a strong threat. The natural resources and the use of heritage are heavily threatened, urging us to contemplate a sustainable development in our actions.

The tourism and heritage relationship is a very sensitive issue and sometimes a cause of disagreements between local and regional actors. One of the main reasons of the dissonance is that tourism and heritage have different perspectives about the use of heritage places. Heritage administrators are increasingly anxious about heritage tourism and accuse it for promoting an accelerating degradation of historical sites, buildings, and other structures. On the other hand, the tourism sector criticizes heritage administration for insufficient efforts to find a viable economic solution for the development and protection of heritage assets.

It should be noted that tourism has been the greatest ally in rehabilitation, conservation, safeguarding natural and cultural heritage in the motivation of communities, but it demands great responsibility from managers, policymakers, and society in general. As will be discussed later, heritage sites and historic towns are prone to overtourism. Tourism can improve conservation and the economic value of cultural heritage and could justify the preservation of cultural heritage and to educate communities and visitors alike. However, if tourism does not respect the intrinsic values of heritage, it can compromise its authenticity and damage its integrity. The development of tourism without particular concern for heritage values endangers the very characteristics of landscapes and the lifestyles of communities that drive it, ultimately degrading the overall visitor experience.

According to the analysis of charts and conventions, risk mitigation can be focused on an action centered on the territorial planning and management, on the protection of the heritage values, and the dissemination of the heritage significance through actions of formal and nonformal education (Fig. 6.2). These actions can promote protection and care for heritage assets, reduce negative impacts, and avoid the destruction of the elements that give value to the territory.

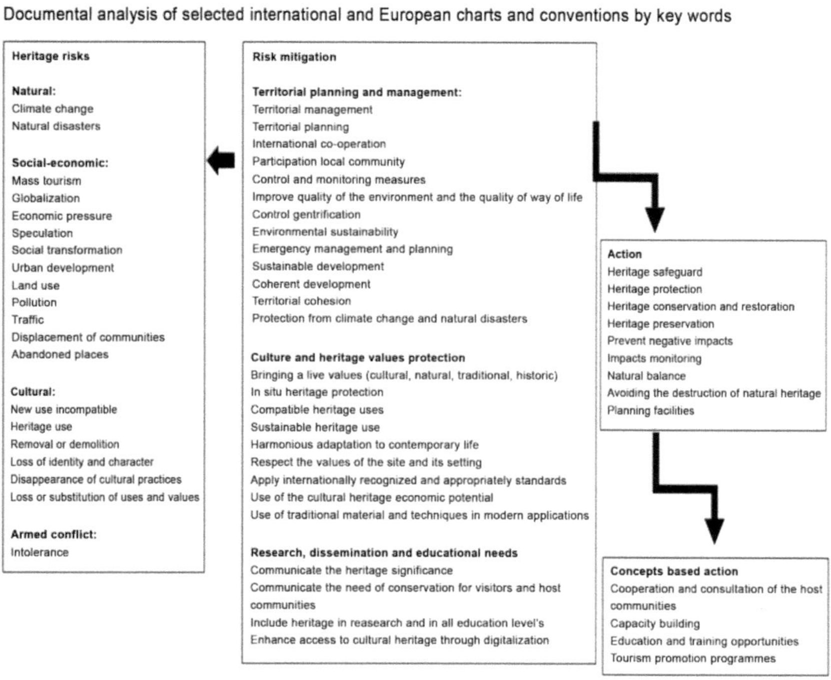

FIGURE 6.2 Risks mitigation.
Source: Diagram prepared by authors.

However, in order to achieve precise objectives in this respect that would benefit the territory and its people, it is important to involve all the stakeholders. They all need to develop their capacity through education and training.

The paramount economic function of tourism—its contribution to the "creation of wealth and employment," as pronounced in the Global Code of Ethics for Tourism (UNWTO, 1999)—is beyond doubt. The steady growth of the international tourism industry's share of the global GDP—amounting to 10% in 2017, following a 5% rise over the previous year (UNWTO, 2018)—is one of the most eagerly quoted economic indicators. It has been observed, time and again, that tourism favors free-market conditions—in particular free enterprise, and the free flow of money, people, and goods, in order to maximize its positive socio-economic effects and to ensure an unencumbered utilization of key tourism assets.

With a short exception at the end of the first decade of the 21st century, the global political and economic conditions have been particularly conducive for tourism and travel to flourish in most of their aspects. Nevertheless, the need to highlight the transformative power of tourism beyond the conventional economic indicators has also been advocated in the literature. Multiple evidence has been presented that tourism has the potential to promote peace, foster solidarity and tolerance, contribute to justice and social equity, enable cultural and intellectual exchange, increase the quality of life, trigger positive social changes, provide rationale for the conservation, protection and interpretation of buildings and the revitalization of traditions, etc. (Higgins-Desbiolles, 2006; Mason, 2016; Rivera et al., 2016; Sharpley, 2014; Smith and Diekmann, 2017). Arguably, the significance of cultural heritage tourism in this respect is greater than that of many other forms of leisure tourism (Council of Europe, 2005; Garrod and Fyall, 2000; McKercher and Du Cros, 2002; Rebanks, 2010; Richards, 2018; Silverman et al., 2017).

However, as a complex, global phenomenon, tourism has not been without its flaws. In fact, the very Code of Ethics was a necessary response to numerous cases of cultural and economic exploitation, social injustice, and environmental destruction caused by the industry around the world. A number of international and sectoral agreements and treaties, some of them discussed in this chapter, have addressed a variety of phenomena that undermine the net positive impact of tourism on people and places. The spectre of overtourism—that is, exceeding the carrying capacity of destinations, mainly those rich in cultural and heritage assets—has been but the latest threat borne out of the global quantitative success of the tourism industry (WTTC, 2017). Moreover, a number of communities and organizations around the world have been marginalized or altogether excluded from participating in the benefits of tourism due to their lack of knowledge, skills, or institutional mechanisms needed to utilize that potential (Silverman et al., 2017; Waterton and Watson, 2011). Digitalization has resolved some of the tensions resident in the discussed phenomena but it also brought about new challenges of exclusion based on digital literacy, as well as the as yet unresolved issue of misinformation (Taylor and Gibson, 2017).

The aim of this section is to identify and characterize the groups of interested or affected parties and/or individuals that have their stake in the tourism industry's deployment and expansion, and whose capacity—or lack thereof—to partake in the above can hinder or enhance the positive net effects of heritage tourism development.

6.3 HERITAGE TOURISM STAKEHOLDERS IN SELECTED LITERATURE

There have been many attempts to identify the social groups, institutions, and other parties that already do or potentially can participate in the sustainable development of cultural and heritage tourism. However, the perception of the role of various groups in this respect has markedly fluctuated since the 1980s (Richards, 2018). It has evolved from a traditional, top-down approach, that assumed the decisive role of experts and government agencies, to the current understanding that the key factor of successful and sustainable cultural tourism development are thriving local communities (Pollock, 2018).

The traditional, authoritative approach to heritage management, also encompassing heritage tourism and its stakeholders still linger in the practice of UNESCO, which, according to Lucas Lixinski, Vice-President of the Association of Cultural Heritage Studies, "falls short of bringing communities into the implementation of its cultural mandate" (Dore, 2018). A similar mindset dominates many intergovernmental and governmental agencies involved in heritage protection and management, and those active in the making and delivery of historical and educational policy, or in the organization and development of tourism (Harrison, 2013; Silverman et al., 2017; Waterton and Watson, 2011). It is marked by what Laurajane Smith famously called the Authorized Heritage Discourse (Smith, 2006), that is, a widespread belief that heritage resources can only be properly interpreted and evaluated by experts, especially at the international level. The concept has been widely discussed within the discipline of Heritage Studies (Ashworth and Howard, 1999; Fredheim and Khalaf, 2016; Silverman et al., 2017; Smith and Campbell, 2015; Throsby, 2001). Many authors point to the limitations of the approach, which has created an authoritative cultural canon of items, sites, and traditions that are protected and preserved at the expense of others, and often not for but against their potential use by various communities (Dore, 2018). Also, it seems to contradict the general trend toward democratization and broadening of the concept of culture itself, and in consequence also of cultural tourism, which has expanded away from traditional elite pursuits to embrace a mass market orientation (Richards, 2018).

Along with the critique of the Authorized Heritage Discourse and an improving understanding of the complex relationship between tourism and heritage, recently it has also been noted that the way various communities engage with cultural and heritage resources has changed over the years (Chabiera et al., 2017; Harrison, 2013; Ripp et al., 2017; Smith et al., 2012). Tourists are no longer seen as impartial, disaffected consumers, and host communities as complacent suppliers of tourism services. In fact, it has been noted that the essence of cultural tourism lies not in the fact of visiting cultural destinations, but in the experience to be had in the process (Smith et al., 2012).

The dominating business perspective in Tourism Studies means that there is scarce literature that would recognize the potential that lies in the encounter between tourists and local communities. In fact, the notion of such an encounter—and its consequences—is primarily an anthropological concept that lies outside the scope of expertise of most economics scholars (Nogués-Pedregal, 2019). Meanwhile, the failure to fully address the dynamics between hosts and visitors, the insiders and outsiders, affects our ability to understand and exploit the phenomenon of culture itself—and to maximize the potential of cultural and heritage tourism.

Thus far, most mainstream contributions on the subject of cultural tourism highlight its inherent dualism, visitors and locals, technology and tradition, protection and use, and profit and significance. The forces that shape the socio-economic processes of tourism seem to be in a continued and mutually competitive strive and the need—and possibility—to find the balance between them has only been uttered very recently (De Wilde, 2018).

As a matter of fact, both tourists and local communities seem to have grown weary of this traditional divide. The tourists' quest of authenticity has evolved from looking for exceptional experiences to the now prevailing search for "the everyday" (Richards, 2018). It is expressed in a growing number of offers to eat or live "like the locals," as well as in the recent rise of community-based or cooperative tourism business models. If not entirely successful, for example, due to deep cultural differences, the trend has at least resulted in the creation of "in-between" experiences in cultural tourism; encompassing elements that do not belong either in the local or in the tourist sphere (Richards, 2018).

At the same time, communities have been increasingly outspoken in their opposition to mass tourism, but the majority of destinations seem to have been unable to provide a comprehensive, scalable alternative. The recent insistence by New Zealand and Hawaii on attracting "better tourists," who would be willing to make real connections with locals—and generate more value to the economy—seem desperate and elitist at the core (Mathews, 2019). What seems to drive the tourist-local divide is a particular misunderstanding surrounding, at the street level, of the value of "the everyday." The antagonism is also often exacerbated by the relations of dependence created by international tourism (Nogués-Pedregal, 2019). At this point, it is worth noting that the transformative power of tourism extends also to places—towns, territories, and landscapes—thus, expanding the list of possible heritage tourism stakeholders to those, who are also direct and indirect users of those places.

In the face of necessity to move beyond the dichotomy of locals and visitors, it is herein proposed that heritage tourism stakeholders are seen as a spectrum, a network, ranging from insiders to outsiders and from the closest heritage communities to those that have the least attachment or involvement with it. Such view corresponds closely with a concept put forth by Ripp (2018), a Regional Coordinator at the Organization of World Heritage Cities, who suggested to approach cultural heritage as a system that principally belongs to local communities. Ripp distinguishes direct heritage users, indirect users, and influential parties—a classification that bypasses questions of power and emphasizes the aspect of heritage use (see Table 6.1). In this perspective local communities—the most immediate heritage users—are an essential element of the heritage system, a key stakeholder; tourists and trades have been classed as secondary users, and the remaining affected or interested institutions are mentioned as a tertiary, but influential, element of the system. Thus, by overturning the current system of power and directing attention toward the use value of heritage, which, at least from an economic point of view, has been the least troublesome of all the possible heritage values discussed in the literature, emerges a new categorization of heritage stakeholders, fitting for a study focused on the practical issue of capacity building.

Indeed, putting people at the center of the concept of heritage was also the core intention of the 2005 Faro Convention on the Value of Cultural Heritage for Society (Council of Europe, 2005). By emphasizing the changing and diverse needs of various heritage stakeholders, and empowering the original heritage users—the local people—the Faro Convention was a major step away from the rigid, linear, and authoritative heritage management approaches of the earlier era. It also emphasized that a sustainable approach demands that heritage continues to play a contemporary role—to be used—for the benefit of its various communities.

Table 6.1 shows the complexity of the heritage system seen this way. It has also been cross-referenced with the classification of heritage stakeholders according to the 2011 UNESCO recommendation on the historic urban landscape, which distinguishes three groups: communities, decision-makers, and professionals (UNESCO, 2011).

6.4 CONCLUSIONS

The approach of conservation and preservation of historical landscape as conservation and preservation of local memories through the symbolic elements of existing heritage is an important instrument for planners to avoid degradation or destruction of the local heritage. The geneses of perennial values are made during a long scale of time, including several successive stages of recognition and selection, projected on a time scale.

In this sense, it is significant to signalize the noteworthy heritage in a territory as a context of living societies in a certain place, and it is essential to create a methodology that can help to identify those important elements needing to be protected. It is imperative in any reflection about landscape and its constructions, to keep the line of thinking of the existent link created between natural and cultural elements.

TABLE 6.1 Heritage Tourism Stakeholders.

	COMMUNITIES		DECISION-MAKERS		PROFESSIONALS
DU	local citizens tourists	INF	policy-makers ministries planners gov't agencies local authorities conservators UNESCO	DU	Museums Monuments site managers historic houses touristic routes
IU	local businesses farmers			IU	DMOs
INF	NGOs educational inst. councillors			INF	Consultants Experts

DU - Direct Users
IU - Indirect Users
INF - Influencial

Source: Author's own compilation.

Every day the action of man shapes the landscape and the heritage resources which are testimonies of the passage of time in result of the human needs and "uses." The local history is changing at the turn of years and centuries and the historical landscape is in constant transformation due to the contemporaneity of human actions. In this context of historical evolution, the relation of man with the surrounding space that implies its transformation is very visible, a fact that could bring degradation or deterioration of the historical resources.

As an important economic resource, heritage must be valued and protected. It is not easy to decide or find alternatives to safeguard heritage, knowing that these cultural resources of the territory acquire enormous value and ample significance because they are active exporters of messages from the past. As elements of the territory that have persisted in the present time, they are living witnesses of the passing of several generations and bearers of a collective memory that we all have as a duty to protect for the future, as the UNESCO (1964) notes from an early date.

Various risks and challenges are constantly facing heritage, including natural disasters, economic development, tourism, pollution, inappropriate management, plunder, and war. It is important to emphasize the issue of economic development which exerts strong pressure on the cultural and natural resources.

At this moment, we are interested in addressing the decisions regarding the deliberation of actors involved in the processes of safeguarding the historical landscape, in particular changes arising from property impacts. It is necessary for the local managers to take into account the risks and pressures of their assets in order to act in accordance with international regulations and safeguard the traces of the past for future generations. In order to do so, it is necessary for local governments and managers to value and communicate this value so that it is known and preserved in the future, according to what it refers to. This is a way of building or developing capacities. Some of the risks bring about an irreversible transformation, both in nature and in the cultural landscape where elements of durability in time produced and produce the identity of cultures and societies.

The importance of heritage and the importance of historical and cultural landscapes to be preserved is not only because of their role as an important evidence to previous times and the source of a spirit of place, but because heritage and culture are important factors of territorial and economic development.

KEYWORDS

- **heritage**
- **tourism**
- **sustainable development**
- **landscape**
- **conservation**

REFERENCES

Ashworth, G. J.; Howard, P. *European Heritage Planning and Management*; Intellect Books: Exeter, 1999.

Australia ICOMOS. *The Burra Charter. The Australia ICOMOS Charter for Places of Cultural Significance*; Australia ICOMOS Inc: Burwood, 1999.

Bourret, C. Cultural Heritage, Tourism, Identity and Sustainable Development of Territories: The Case of the Compostela Ways. In *International Multidisciplinary Scientific Conference on Social Sciences & Arts*. SGEM, 2017.

Carroll, P.; Aarrevaara, E. Review of Potential Risk Factors of Cultural Heritage Sites and Initial Modelling for Adaptation to Climate Change. *Geosciences* **2018,** *8* (9), 322.

Chabiera, A.; Dąbrowski, A.; Fortuna-Marek, A.; Kozioł, A.; Lubaś, M.; Nowak, P.; ... Stępnik, K. *Polacy wobec dziedzictwa. Raport z badań społecznych*; Warszawa, 2017.

Council of Europe. Council of Europe Framework Convention on the Value of Cultural Heritage for Society, 2005. http://conventions.coe.int/Treaty/EN/Treaties/html/199.htm.

Dans, E. P.; Gonzalez, P. A. The Altamira Controversy: Assessing the Economic Impact of a World Heritage site for Planning and Tourism Management. *J. Cult. Heritage* **2018,** *30*, 180–189.

De Wilde, P., Ed. *Travel to Tomorrow*; VISITFLANDERS: Brussels, 2018.

Dore, B. Who Controls Cultural Heritage?, 2018. https://hyperallergic.com/477859/who-controls-cultural-heritage/ (accessed Feb 16, 2019).

Feria, J., Ed. *Territorial Heritage and Development*; London: CRC Press, 2012.

Francis-Lindsay, J. The Intrinsic Value of Cultural Heritage and Its Relationship to Sustainable Tourism Development: The Contrasting Experiences of Jamaica and Japan. *Caribb. Q.* **2009,** *55* (2), 151–168.

Fredheim, L. H.; Khalaf, M. The Significance of Values: Heritage Value Typologies Re-examined. *Int. J. Heritage Stud.* **2016,** *7258* (April). https://doi.org/10.1080/13527258.2016.1171247.

Freitas, I.; Cortés, R. E.; Leite, P. P. A Methodological Proposal for Building Conservation: A Case in Guimarães, World Heritage City. In *Conservation and Promotion of Heritage Tourism*, 2018; pp. 57–86.

Freitas, I.; Martín Cea, J. C.; Villanueva Zubizarreta, O.; Val Valdivieso, I. Abordagem metodológica de conservação da história e do patrimônio: O Vale do Douro. *Revista Tempo e Argumento* **2017,** *9* (21), 99–125.

Garrod, B.; Fyall, A. Managing Heritage Tourism. *Ann. Tour. Res.* **2000,** *27* (3), 682–708. https://doi.org/10.1016/S0160-7383(99)00094-8.

Guzmán, P. C., Pereira Roders, A.; Colenbrander, B. J. (2017). Measuring Links between Cultural Heritage Management and Sustainable Urban Development: An Overview of Global Monitoring Tools. *Cities* **2017,** *60* (A), 192–201.

Harrison, R. *Heritage. Critical Approaches*; Routledge: London, 2013.

Helly, D. Scenario Configurations for the EU as a Cultural Heritage Protection Actor in Armed Conflicts. *J. Cult. Heritage* **2018,** *33*, 271–277.

Higgins-Desbiolles, F. More than an "industry": The Forgotten Power of Tourism as a Social Force. *Tour. Manage.* **2006,** *27* (6), 1192–1208. https://doi.org/10.1016/j.tourman.2005.05.020.

ICOMOS. *The La Valletta Principles for the Safeguarding and Management of Historic Cities, Towns and Urban Areas*; La Valletta, 2011.

ICOMOS. *The Australia ICOMOS Charter for Places of Cultural Significance, the Burra Charter; Australia, 2013.*

Koskowski, M. R. Castles as a Particular Tourist Product. *Forum UNESCO University and Heritage, 10th International Seminar Cultural Landscapes in the 21st Century*; Newcastle upon Tyne, 2006; pp. 11–16. http://conferences.ncl.ac.uk/.

Koskowski, M. R. Dziedzictwo kulturowe w przestrzeni turystycznej na przykładzie zamków w Polsce. In *Ożywianie przestrzeni turystyki kulturowej*; Ostrowski, D., Ed.; Wyższa Szkoła Turystyki i Języków Obcych: Warszawa, 2017; pp. 69–83.

Kourilova, J.; Pelucha, M. Economic and Social Impacts of Promoting Cultural Heritage Protection by the Czech Rural Development Programme 2007–2013. *Eur. Countryside* **2017,** *9* (3), 486–503.

Lwoga, N. B. Dilemma of Local Socio-economic Perspectives in Management of Historic Ruins in Kilwa Kisiwani World Heritage Site, Tanzania. *Int. J. Heritage Stud.* **2018,** *24* (10), 1019–1037.

Mason, P. *Tourism Impacts, Planning and Management*; Routledge: London, 2016.

Mathews, J. Hawaii Doesn't Need More Tourists, It Needs Better Tourists. *Zócalo*, 2019. http://www.zocalopublicsquare.org/2019/02/22/hawaii-doesnt-need-tourists-needs-better-tourists/events/the-takeaway/.

McKercher, B.; Du Cros, H. *Cultural Tourism : The Partnership between Tourism and Cultural Heritage Management*; Haworth Hospitality Press: London, 2002. https://books.google.pl/books?id=9FxsBgAAQBAJ.

Nogués-Pedregal, A.-M. Anthropological Contributions to Tourism Studies. *Ann. Tour. Res.* **2019,** *75*, 227–237.

Petroman, C.; Iancu, T.; Popescu, G.; Iosim, I.; Csaholczi, A.; Marin, D.; Vaduva, L. Directed Authenticity in Tourism. *LUCRĂRI ŞTIINŢIFICE* **n.d.**, *20* (1), 172–176.

Pierantoni, L. Governing Regional Development through Culture and Creativity: The Case of the Veneto Region. *Eur. Plan. Stud.* **2015,** *23* (5), 963–978.

Pollock, A. Towards a Flourishing Destination. In *Travel to Tomorrow*; De Wilde, P., Ed.; VISITFLANDERS: Brussels, 2018; pp. 16–17.

Rebanks, J. The Price of Lemons—World Heritage and Economic Development. Interview with James Rebanks. *World Heritage* **2010,** *58*, 79–82. www.pfdheritage.com/wh58.

Richards, G. Cultural Tourism: A Review of Recent Research and Trends. *J. Hosp. Tour. Manage.* **2018,** *36*, 12–21. https://doi.org/10.1016/j.jhtm.2018.03.005.

Ripp, M. Heritage as a System and Process that Belongs to Local Communities. Reframing the Role of Local Communities and Stakeholders, 2018. https://www.researchgate.net/publication/325398758_Heritage_as_a_System_and_Process_that_Belongs_to_Local_Communities_Reframing_the_role_of_local_communities_and_stakeholders.

Ripp, M.; Goettler, M.; Scheffler, N., Eds. *Community Involvement in Heritage Management*; Organisation of World Heritage Cities: Regensbur, 2017. https://www.researchgate.net/publication/318531316_Community_Involvement_in_Heritage_Management_-_Guide-book.

Rivera, M.; Croes, R.; Lee, S. H. Tourism Development and Happiness: A Residents' Perspective. *J. Dest. Market. Manage.* **2016,** *5* (1). https://doi.org/10.1016/j.jdmm.2015.04.002.

Scheyvens, R. *Tourism for Development: Empowering Communities*; Prentice Hall: Harlow, 2002.
Seminara, G.; Sgamellotti, A. Resilience of Cultural Heritage to Natural Disasters: The Recent Involvement of Science Academies. *Abitare La Terra* **2017,** *16* (42/43), 47–49.
Sharpley, R. Host Perceptions of Tourism: A Review of the Research. *Tour. Manage.* **2014,** *42*, 37–49. https://doi.org/10.1016/j.tourman.2013.10.007.
Silverman, H.; Waterton, E.; Watson, S., Eds. *Heritage in Action*; Springer International Publishing: Cham, 2017. https://doi.org/10.1007/978-3-319-42870-3.
Smith, L. *Uses of Heritage*; Routledge: Abingdon, 2006.
Smith, L.; Campbell, G. *The Tautology of "Intangible values."* Australian National University: Canberra, 2015. http://www.canberra.edu.au/research/faculty-research-centres/cccr/resources/insignificance
Smith, L.; Waterton, E.; Watson, S., Eds. *The Cultural Moment in Tourism*; Routledge: London, 2012.
Smith, M. K.; Diekmann, A. Tourism and Wellbeing. *Ann. Tour. Res.* **2017,** *66*, 1–13. https://doi.org/10.1016/j.annals.2017.05.006.
Taylor, J.; Gibson, L. K. Digitisation, Digital Interaction and Social Media: Embedded Barriers to Democratic Heritage. *Int. J. Heritage Stud.* **2017,** *23* (5), 408–420.
Throsby, D. *Economics and Culture*. Cambridge University Press: Cambridge, 2001.
UNESCO. *The Venice Charter for the Conservation and Restoration of Monuments and Sites*; Venice, 1964.
UNESCO. *Convention Concerning the Protection of the World Cultural and Natural Heritage*; Paris, 1972.
UNESCO. *Convention for the Safeguarding of the Intangible Cultural Heritage*; Paris, 2003.
UNESCO. *Québec Declaration on the Preservation of the Spirit of Place*; Québec, 2008.
UNESCO. *Recommendation on the Historic Urban Landscape*; Paris, 2011.
UNWTO. *Global Code of Ethics for Tourism*; Santiago, 1999. http://cf.cdn.unwto.org/sites/all/files/docpdf/gcetbrochureglobalcodeen.pdf.
UNWTO. *Tourism Highlights 2018*; Madrid, 2018.
Vecco, M.; Imperiale, F. Cultural Heritage: Values and Measures: What Insurance Value? *J. Multidiscip. Res.* **2017,** *9* (1), 7–22.
Vinuesa, M. A.; Torralba, L. Visión territorial del patrimonio y sostenibilidad del turismo. *Boletín de La Asociación de Geógrafos Españoles* **2018,** *78*, 212–244.
Waterton, E.; Watson, S., Eds. *Heritage and Community Engagement. Collaboration or Contestation?* Routledge: London, 2011.
WTTC. *Coping with Success. Managing Overcrowding in Tourism Destinations*, 2017. https://www.mckinsey.com/industries/travel-transport-and-logistics/our-insights/coping-with-success-managing-overcrowding-in-tourism-destinations.
Zhu, Y. Uses of the Past: Negotiating Heritage in Xi'an. *Int. J. Heritage Stud.* **2018,** *24* (2), 181–192.

CHAPTER 7

PUBLIC–PRIVATE PARTNERSHIP: A PROBABLE PANACEA FOR ECOTOURISM DEVELOPMENT IN ANAMBRA STATE, NIGERIA

ODUM CHIGOZIE JUDE[1*], F. O. F. ONWUDUFOR[2], and UCHENNA NGOZI ARENE[1]

[1]*Department of Archaeology and Tourism, University of Nigeria, Nsukka, Enugu, Nigeria*

[2]*Department of History and International Relations, Nwafor Orizu College of Education, Nsugbe, Anambra, Nigeria*

[*]*Corresponding author. E-mail: jude.odum@unn.edu.ng*

ABSTRACT

Why is tourism not developed in Anambra state, despite enormous tourist attractions? Who manages tourist attractions in Anambra state? Public–private partnerships (PPPs) seem to offer a lot of hopes in providing efficiency where government management is not good enough. Some African nations have turned to PPP to solve troublesome developmental issues. Although some scholars have accused PPP of coming with mixed benefits, but, it ensures productive output and increases in infrastructural development. Unfortunately, there is paucity of literature on PPP and ecotourism development in developing nations like Nigeria. Therefore, the study aims to know the possibility of using PPP to develop ecotourism attractions in southeastern Nigeria, using Anambra State as reference point. The study adopted both quantitative and qualitative method. The former entails selection of key informants in selected communities, namely, Amaokpala, Agulu, Owerrezukalla, Ufuma, Ogbunike, Awka, Okpeze, Ndiukwuenu, and staff of related ministries. The quantitative aspect entails using a total of 480 questionnaires distributed randomly across these communities. The study revealed that most of the ecotourism attractions in Anambra State are not developed due to meager state resources, shortage of tourism experts, and government poor attention to tourism sector tourism sector. Furthermore, these communities accepted that PPP is the best strategy to develop natural attractions in their communities; this was collaborated by results from related ministries. The chapter thus argues that to develop ecotourism attractions in Anambra State, PPP is one of the methods that will be used and accepted by these communities; second, the host communities believe that it will ensure sustainable development, whereas staff of related ministries are of the view that it will augment government poor attention and funding to tourism development in the state. Therefore, the PPP agreement must be economically beneficial, culturally acceptable, and environmentally compatible to ensure sustainability. The study is of immense benefit to tourism stakeholders, policymakers, and invaluable to the academic world.

7.1 INTRODUCTION

Anambra State has hundreds of tourist attractions existing in different forms: natural, manmade, historic, and cultural. Developing tourist attractions in the state has been sluggish, and most of the attractions are not known beyond the community and local government where they exist. This has made the touristic appeal of these attractions very poor (Odum, 2017). Meanwhile, some of these tourist attractions despite being undeveloped generate little income to the host communities endowed with them, especially eco-tourist attractions like caves. A well-developed eco-tourist attraction has the capacity to generate income for the state and host communities, create employment for the locals, and help to conserve the attraction itself for posterity by using funds generated from visitors and tourists, while its academic and scientific merits cannot be overlooked.

Anambra State is blessed with many eco-tourist attractions such as streams, lakes, rivers that harbor aquatic organisms ranging from different species of fishes to crocodiles. There are caves with beautiful scenery and tranquil environment; moreover, some of these caves harbor some fauna that adds beauty to them, like lions (*Panthera leo*), grass-cutter (*Thryonomys swinderianus*), porcupine (*Erethizon dorsatum*) (Onwudufor and Odum, 2014), and beautiful vegetation. The state has about eight forest reserves and many ungazetted forests; many of the ungazetted forests harbor fauna such as *Cercopithecus sclateri*, a species of monkeys found near forests in Umuawulu axis. This species of monkey is adjudged to be endemic in the area (Baker, 2012 in Odum, 2014) with attendant environmental values.

Therefore, the chapter aims at exploring public–private partnership (PPP) as a way of developing eco-tourist attractions especially eco-tourist attractions in the state given that PPP is usually advocated due to quest for infrastructure, paucity of government fund, and the willingness of the private sector to assist (Herpen, 2002). The limited number of professionals to manage and develop tourism resources in has been a challenge facing tourism in the state (Odum, 2018). PPP is advocated for where the government can state the quality of service they need from the private sector, and the service measured against specified indicators. And, an effective PPP ensures high-quality service delivery to consumers and government at cheaper cost unlike government provision of such services (International Monetary Fund, 2004).

What are the eco-tourist attractions in Anambra state? Why are eco-tourist attractions in Anambra state undeveloped? Who is responsible for managing these attractions? Can PPP help in developing these attractions? What is the view of community people and government staff of ecotourism-related ministries about using PPP to develop ecotourism resources? Given this background, the study is set to highlight some eco-tourist attractions in Anambra state and their current status, evaluate perceptions of selected communities and staff of Ministry of Diaspora Affairs, Culture, and Tourism and Ministry of Environment regarding the use of PPP in developing tourism in the state.

This study is relevant because most tourist attractions in the state are undeveloped; therefore, exploring opportunities for developing these attractions is paramount due to the need to diverse the state and Nigeria source of income. Besides, tourism seems to offer an alternative to the nation's over reliance on crude oil (Okonkwo and Odum, 2010). Second, getting the perception of community members on using PPP as an approach to develop eco-tourist attractions is necessary as the findings will be relevant to policymakers who need them to avoid formulating policies that are not only in tandem with the host community realities but also ones that may jeopardize the interest of these attractions. Furthermore, this seems to be the first attempt to X-ray PPP as an option for ecotourism development in Anambra State.

7.1.1 LITERATURE REVIEW

7.1.1.1 ECOTOURISM ATTRACTIONS

Nigeria is endowed with a lot of ecotourism attractions. Odunlami (2003) has assessed the ecotourism potential of Yankari National Park. Ijeomah and Alao (2007) did a study on the popular Farin Ruwa in Nasarawa state, and Ajake et al. (2013) did a related study in an uncharted natural attraction in Cross River State.

Within the southeastern part of Nigeria, Ewelum (2015) carried out an assessment of ecotourism attractions in Enugu state. Ecotourism attractions in Anambra State have been studied by different scholars; some of the studies that are tourism related are Agulu lake (Aziekwe, 2008; Okeke and Nwokolo, 2008; Odum, 2017), Amaokpala lake (Odum, 2017), Ogbunike cave (Okani, 1996; Odum, 2011, 2017), Owerrezukalla cave (Itanyi et al., 2013), Ajalli cave (Oguamanam and Nwankwo, 2015), and Ufuma cave (Onwudufor and Odum, 2015; Odum, 2017). Ukpaka (2014) highlighted the implications of surviving flora in Anambra state; Odum (2017) studied Umuawulu forest (ungazetted) and gazetted forest reserves, namely, Akpaka Forest Reserve, Agu-aba Forest Reserve, and Mamu forest.

All the studies cited above directed the attention of government and community members to the benefits (income, jobs, and conservation) of developing these ecotourism attractions, their challenges and prospects, as well as their academic merits. Nevertheless, none of these studies explored the role of PPP in developing these attractions. Although Odum (2017) hinted at PPP as a way of developing these attractions, he failed to give a detailed account of how it could be done.

7.2 THE RATIONALE FOR PPP IN TOURISM

Economic meltdown has made governments to look for alternative means of funding projects and privatization of some services that it usually provides (Hall, 1999, 2008), and tourism, being one of the sectors that is capital intensive, has necessitated the need for it to be funded through any other means other than government. This is further enhanced by the interdependent nature of tourism sector leading to the call for interorganizational relationship. Discourse on collaboration and partnership has elicited different approaches and perspectives, leading to terms like community-based tourism (Maria and Hall, 2012). It has been adjudged to be beneficial in tourism development. Collaboration in developing tourist attraction seems to be envogue and one of such collaborations is PPP.

7.2.1 PUBLIC–PRIVATE PARTNERSHIP

Goloran et al. (2015) have shown where PPP collaboration was used in promoting in World Heritage Site and drawn extra resources for training of locals and building infrastructure in Puerto Princesa Subterranean River National Park in Philippines. It has helped the locals in getting more skills as tourist guides, generate income, and protection of invaluable cultural heritage in Botswana, the Tsodilo Hills World Heritage Site (Mbaiwa, 2015). PPP can equally be all encompassing by involving both national and regional government with international organizations or agencies. This was demonstrated in Yangon and Nyaungshwe Lakes of Myanmar (located in geographic boundaries of China and India).

The government of Tanzania using PPP was able to restore and conserve the Chumbe Island located off the coast of Zanzibar. Environmental nongovernmental organization was drafted to sensitize the people and engage local fishermen to assist in the protection of the native ecosystem that has coral reef under threat due to overfishing and unwholesome fishing practices. The destination was revived and maintained without any cost from the government.

Tanzina et al. (2013) listed countries where PPP has been of immense benefit to such as Russian government soliciting for PPP for 2014 Winter Olympics; India with successful record of PPP in Jal Mahal Tourism Projects. Within the same country, PPP was adjudged to be successful in Kerala Travel Mart (Menon and Edward, 2014). It was used successfully in the London Olympic Games in 2012. South Africa has some of its National Parks contracted out for management through PPP; one of the parks is Kruger National Park. Ten years' build-operate-transfer (BOT) approach was used (Peter, 2005). Collaboration has been seen to be useful in development of tourism (Menon and Edward, 2014; Kumar et al., 2015; Ekpenyong and Mmom, 2015), therefore keying into collaboration like PPP will be ideal for ecotourism development resources in Anambra State.

7.3 ANAMBRA STATE

Anambra State is geographically located in southeastern Nigeria and lies between longitudes 6°35′E; 7°30′ and latitudes 5°40′N; 6°48′N. The state is bounded in the west by Delta state, south by Imo state, north by Enugu, and in the northwest by Kogi. The state is made up of 21 local government areas namely: Awka North, Awka South, Aguata, Anambra East, Anambra West, Anaocha, Ayamelum, Ekwusigo, Dunukofia, Idemili North, Idemili South, Ihiala, Njikoka, Nnewi North, Nnewi South, Ogbaru, Onitsha North, Onitsha South, Oyi, and Orumba North and South. Onitsha, Awka, and Nnewi are the major cities in the state (Odum, 2017) with Awka as the state capital and seat of the government administration. The state witnessed political instability since the return of democracy in in Nigeria in 1999. Rigging of election and court cases have made most of the governors not to complete their usual 4-year tenure until in the recent past. Such political instability and inconsistency cannot aid any meaningful developmental plan for ecotourism in the state.

The flora characterizing Anambra State are as follows: Oil palm tree (*Elaeis guineenisis*), breadfruit (*Treculia africana*), mangoes (*Mangifera indica*), coconut (*Cocos nucifera*), African star apple (*Chrysophyllum albidium*), Africa pear (*Dacroydes edulis*), oranges (*Citrus sinensis*), oil bean (*Pentachlethra macrophylla*), *Newbouldia leavis* (*Ogirisi*), and so on. Most of these trees are of economic value to the people of the state. It is a major source of income for some families. Anambra has the following forest reserves: Achalla Forest Reserve with a size of 202.5 ha, Mamu Forest Reserve with 8183 ha, Anambra East Forest Reserve with 1457 ha, Akpaka Forest Reserve with 109.02 (decimated), Osomari Forest Reserve with 12,098 ha, Aguaba Forest Reserve with 222.63 ha, and Nkachu-Ituku Forest Reserve with 108.78 ha (Odum, 2017). In the state, there are other forest relics in the following towns: Umuawulu, Nzam, Orumbanassa, Igbokaenyi, Ukwulu, and so on. These are not gazetted, and presently most of them are being threatened due to population increase and quest for residential houses (Odum, 2017).

7.4 METHOD

Mixed method of research was used in this study. The qualitative aspect comprises in-depth interviews with principal informants within the communities and Department of Tourism, Ministry of Diaspora, Culture and Tourism, and Forestry Department of Ministry of Environment. The Director of Tourism in the Ministry of Diaspora, Culture and Tourism and the Director of Forestry in Ministry of Environment were interviewed and other staff of the ministries who have spent over 10 years in the civil service of the state. Key informants were selected in the eight communities based on age and experience, position/post in the community like town union presidents, relationship with the attractions, especially chief priests of lakes/caves. The key informants comprise kings, cave guards, chief priests, Director of Forestry in Anambra state, the state zonal forest officers, and so on. The quantitative aspect entails distribution of questionnaires randomly in the communities and within the selected ministries (Ministry of Diaspora Affairs, Culture and Tourism, and Ministry of Environment). Based on the number of staff in the aforewritten ministries, an arbitrary sample of 40 questionnaires was shared equally in both ministries within relevant departments in each ministry, namely, Tourism Department in Ministry of Diaspora, Culture and Tourism, and Department of Forestry in Ministry of Environment and 20 questionnaires were administered in each department. The Taro Yamane (1967) formula was applied to generate the sample size for the study. The population of each of the communities is used as a basis for determining the minimum sample size required in Table 7.1. The formula is given as:

$$n = \frac{N}{1 + N(e)^2}$$

where n is the sample size; N is the finite population; e is the level of significance (0.05 level of significance); and 1 is the unity.

The population of each of the communities was projected from the 1991 figure to 2014 given a population growth of 2.21% (National Population Commission Office in Awka) per annum, using the following formula:

$$P = P_n\left(1 + r\right)^n$$

where P is the projected population; P_n is the population of a given year (here, the P_n is 1991); n is the number of years (1991–2014, i.e., 23 years); and 1 is the unity (a constant); and r is the growth rate.

Therefore, in applying the Yammne (1967) formula:

$$\frac{N}{1 + N(e)^2}$$

Substituting,

$$\frac{164{,}741}{1+164{,}741\,(0.05)^2}$$

$$n = 400$$

A total of 480 copies of questionnaires were distributed across the eight communities and the number allotted to each community is given in Table 7.2. Taro Yamane formula gave 400 as the minimum sample size. It is assumed that out of every 10 questionnaires, 2 may not be returned as a result of what the researcher observed during reconnaissance surveys, such as low turnout and peoples lack of interest in study of this nature with fear that government wants to know about them, less population in some of the sampled communities that will make the data generated to be too scanty. Therefore, additional 20% was added to the sample size to make it 480.

TABLE 7.1 Projected Population from 1991 to 2014.

S/No.	Communities	Population as 1991	Projected to 2014
1	Agulu	49,310	81,524
2	Ufuma	24,202	40,013
3	Amaokpala	3182	5261
4	Awka	58,225	96,312
5	Ogbunike	19,985	33,041
6	Okpeze	1834	3032
7	Owerre-Ezukala	6027	9964
8	Ndiukwuenu	1976	3267
9	Total	**164,741**	**272,414**

Source: Authors' computation.

Secondary data were used to support the primary data. Articles relating to PPP were consulted and they helped to shed light on the variants of PPP that are in literature.

It is worthy to note, the study lasted for three years due to financial constraints and verifying the status of these attractions was necessary. This entailed ground trothing and helped in creating Table 7.3 and location of key features with global positioning system, which aided in the creation of the map with selected communities in the study area. The study is limited to selected attractions in the following communities: Agulu Lake in Agulu, Iyi-ocha in Amaokpala, Ogba Ogbunike in Ogbunike, Ufuma cave in Ufuma, Ogba-ukwu in Owerrezukalla, Mamu Forest Reserve within Ndiukwuenu and Okpeze, and Agu-aba Forest Reserve in Awka as highlighted in Figure 7.1.

TABLE 7.2 Distribution of Questionnaire.

Communities	Projected population (2014)	Proportionate sample of questionnaire per community	Questionnaires per community (20%)	Questionnaires returned
Agulu	81,524	120	144	140
Ufuma	40,013	59	71	68
Amaokpala	5261	8	9	9
Awka	96,312	141	169	165
Ogbunike	33,041	49	59	53
Okpeze	3032	4	5	5
Owerre-Ezukala	9964	14	17	18
Ndiukwuenu	3267	5	6	6
Total	**272,414**	**400**	**480**	**464**

Source: Authors' computation.

7.5 FINDINGS

There are innumerable tourist attractions in Anambra state. Some of these attractions are listed in Table 7.3 while the current status is explored in the column named remarks.

Forest reserves in the state are managed by the Ministry of Environment. The number of gazetted forest reserves is eight. Akpaka forest has been decimated, replaced with a housing estate. Agu-aba Forest Reserve is currently threatened by development. These forest reserves are poorly managed due to the following: poor government attention, inadequate forest officers to manage these reserves, community/indigenes encroachment, weak regulation, lack of forest policy in the state, use of outdated forest laws, and so on. While government-owned forest reserves are saddled with these challenges, the ungazetted ones are heavily logged, and most of the communities have relied on these ungazetted ones to make a living. Hunters make a living from these ungazetted forests: a reality that has caused some animals in the forest to relocate, especially *C. sclateri*, which the hunters in Umuawulu axis attested that they rarely see nowadays. This is also partly engendered by other anthropogenic disturbances. For instance, community members harvest nonforest timber products (NTFPs) like bush mango (*Irvingia gabonensis*) from these reserves. It is worthy of note, these ungazetted forests have more diversity of flora and fauna than the gazetted reserves that is filled with Gmelina (*Gmelina arborea*) and teak (*Tectona grandis*) conserved mainly for wood production.

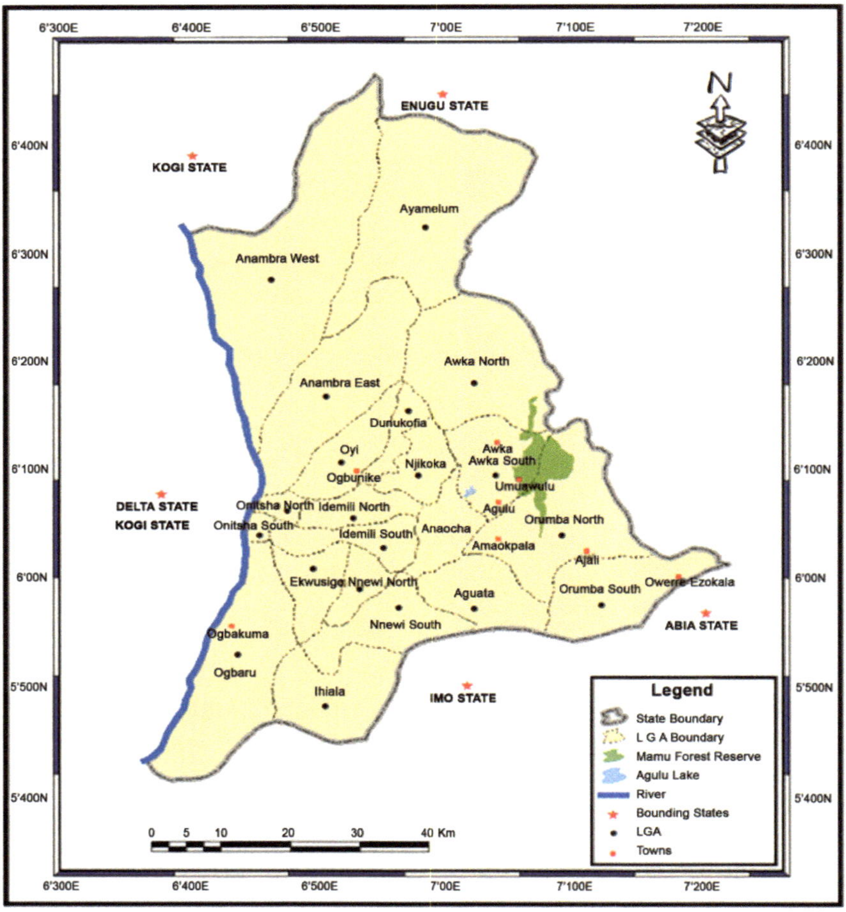

FIGURE 7.1 Anambra state showing the sampled towns in red dots.
Source: GIS laboratory, Department of Geography, University of Nigeria, Nsukka.

TABLE 7.3 List of Some Eco-tourist Attractions in Anambra State.

S/no.	Tourist attraction	Type of attraction	Location	Ownership	Remarks
Caves					
1	Ogba Ogbunike	Natural	Ogbunike	Community	Attracted UNESCO attention and state government attention. For example, state government has built a road leading to it
2	Ogba Ajalli	Natural	Ajalli	Community	Managed by community; not developed
3	Ogba Ufuma	Natural	Ufuma	Community	Managed by a family; not developed and rarely known
4	Ogba ukwu	Natural	Owerrezukalla	Community	Managed by the community; state government has shown interest in its development but no physical proof yet
5	Ngene-Oji	Natural	Mbaukwu	Community	Not developed
Lakes and rivers					
6	Ngene-oji	Natural	Mbaukwu	Community	Not developed. Unknown
7	Agulu Lake	Natural	Agulu	Community	Managed by community; government has built a hotel close to the lake. Filled with crocodiles that are believed to be sacred
8	Iyi-ocha	Natural	Amaokpala	Community	Managed by community; not developed
9	Obutu Lake	Natural	Omogho	Community	Not developed
10	Nawfia Lake	Natural	Nawfia	Community	Not developed
11	Otumoye Lake	Natural	Onitsha	Community	Not developed; polluted, affected by urbanization
12	Ajiogbe Lake	Natural	Aguleri-out	Community	Not developed

TABLE 7.3 *(Continued)*

S/no.	Tourist attraction	Type of attraction	Location	Ownership	Remarks
13	Obizi River	Natural	Nise/Mbaukwu/Nibo, etc.	–	Not harnessed for any tourism purpose
14	Nkisi River	Natural	Onitsha/Nkpor	–	Not harnessed for any tourism purpose
15	Ezegu River	Natural	Ufuma	Community	Not developed. The bank is used for rice plantation
Forests					
16	Achalla Forest Reserve	Natural	Achalla	Government	Gazetted; poorly managed
17	Mamu Forest Reserve	Natural	Okpeze/Ndiukwuenu	Government	Gazetted; poorly managed
18	Anambra East Forest Reserve	Natural			Gazetted; poorly managed
19	Akpaka Forest Reserve	Natural	Onitsha	Government	Gazetted; poorly managed
20	Ogbakuba Forest Reserve	Natural	Ogbakuba	Government	Gazetted; poorly managed
21	Agu-aba Forest Reserve	Natural	Awka	Government	Gazetted; poorly managed
22	Nkachu-Ituku Reserve	Natural		Government	Gazetted; poorly managed
23	Forests in Umuawulu	Natural	Umuawulu	Community	Ungazetted forest; filled with different flora and fauna; like the endemic *Cercopithecus sclateri* is found in the forest
24	Forest in Ukwulu	Natural	Ukwulu	Community	Ungazetted filled with different species of flora

Source: Authors' fieldwork (2017).

Streams, lakes, and rivers in the state are not developed for tourism purposes. These attractions are currently managed by host communities. Communities around these water bodies use these resources to make a living through fishing, while others cultivate rice at the bank of their river like in Ozegu river in Ufuma; others use their for spiritual purposes like Agulu Lake. The Lake is known to have different species of aquatic organisms especially crocodiles. These crocodiles are totem animals and they hold a lot of unique appeal for tourists/visitors and have made the lake popular.

Caves of different sizes and geologic formations exist in the state. These caves are managed by the community within their capacity, with sole interest in the economic benefits associated with visitors. Visitors, researchers, and tourists are charged with different amounts of money to access these caves. The most popular among them is Ogbunike cave which has attracted the attention of UNESCO and state government. The state has built a public convenience near the cave to control open defecation.

Caves, streams, lakes, rivers, and forests in the state have not been developed for tourism purposes. This is caused by little government attention, poor documentation of tourist attractions in the state, inadequate manpower, and poor public knowledge about these attractions.

It is the responsibility of government to harness and manage these eco-tourist attractions but this has failed over the years. Communities have resorted to managing them, and this has resulted in variation in prices to enjoy some of these attractions especially the caves, where some communities charge up to a thousand naira per (between 2-3 dollars) person. Meanwhile, there is no stipulated price anywhere for the intending visitor, tourist, or researcher to know what price s/he is to pay on the site. The money realized goes into the community purse. However, these communities do not have the financial capability to develop these attractions. Besides, the community management at the moment is poor as they are interested only in the money they generate from visitors. Unfortunately, unregulated and uncoordinated visits to these caves have led to their defacement. For instance, as regards to Ogbunike cave, the environment is often littered with bowls and items used for sacrifice in worshiping the deity the cave is associated with. Moreover, forest reserves are not well managed let alone developing them for tourism interest.

7.6 STATISTICAL ANALYSIS

A total of 25 respondents from staff of the related ministries used in the study form the bulk of the respondents. A total of 40 copies of questionnaires were distributed in both ministries with each having 20 copies. After a month's interval, 25 copies of the questionnaire were returned from both ministries. Tourism department has 14 respondents representing 56%, while forestry department has 11 respondents representing 44% as represented in Figure 7.2.

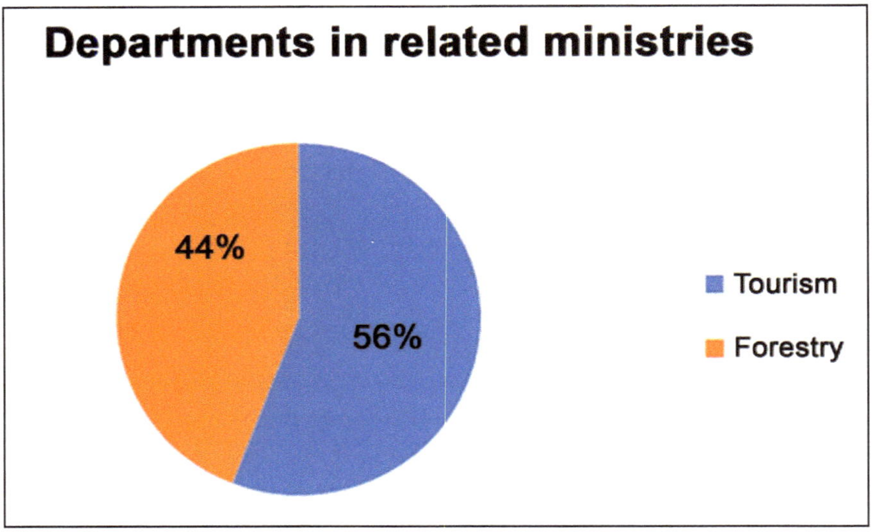

FIGURE 7.2 Ministries used in the study.
Source: Authors' fieldwork (2017).

Build-operate-own seems to be a better option for developing eco-tourist attractions in Anambra State for the following reasons: the tenured nature of government which might affect BOT option because a new government might not be interested in developing tourism in the state; rather, attention will be given to other infrastructure the electorate can see before the end of 4 years in office, which has always been the case over the years; second, the state seems not to have enough fund to invest in tourism since tourism is capital intensive, while build-operate-own will give the managers opportunity to employ best hands instead of government employment that is usually marred with bias and corrupt practices, coupled with the attitude of government employees who see government work as "our things" and that government job is not to be taken seriously. However, build-operate-own management will make sure that profit margin is gotten to ensure prompt payment of salaries of its workers. Moreover, it must remit an agreed percentage to the government coffers per annum on an agreed amount, which can be reviewed periodically.

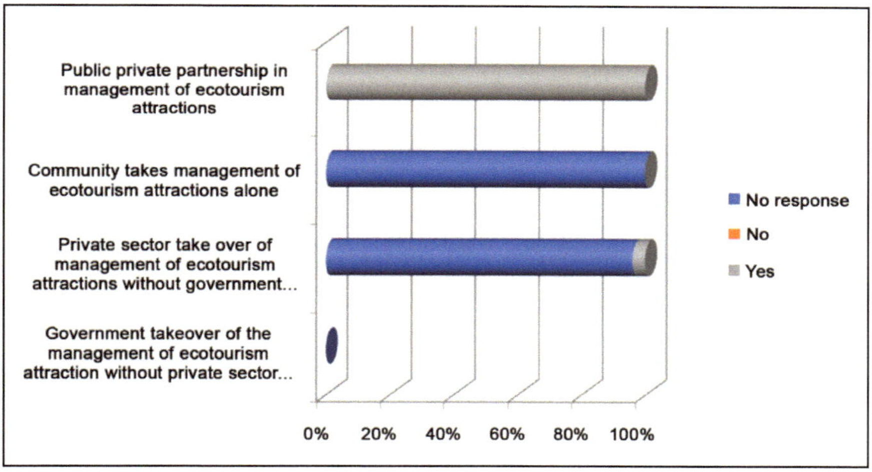

FIGURE 7.3 Categorization of development options from staff of the Department of Tourism and Forestry of the Ministries used for the study.
Source: Author's fieldwork (2017).

7.7 DISCUSSION OF FINDINGS

Table 7.3 shows a list of some of the ecotourism attractions in Anambra state. Ninety-five percent of these natural attractions are not known or developed. Some of them are not known beyond the immediate community where they are located, while the government agency that is meant to document and manage these attractions does not have a comprehensive list of tourist attractions in the state (Odum, 2017), although this may not be unconnected to government poor funding of the tourism department in the State. Table 7.3 reveals that most of these attractions are not developed. This was part of what necessitated this study and the call for alternative means of developing them other than government funding and development. Besides, host communities are yearning for the development of these resources over the years and they see government as not being sincere toward developing tourism resources.

TABLE 7.4 Variants of Public–Private Partnership.

S/No.	Schemes	Modalities
1.	BOO BDO DCMF	The private sector designs, builds, owns, develops, operates, and manages an asset with no obligation to transfer ownership to the government. These are variants of DBFO schemes
2.	BBO LDO WAA	The private sector buys or leases an existing asset from the government, renovates, modernizes, and/or expands it, and then operates the assets again with no obligation to transfer ownership back to the government
3.	BOT BOOT BROT BLOT BTO	The private sector designs and builds an asset, operates it, and then transfers it to the government when the operating contract ends or at some other prespecified time. The private partner may subsequently rent or lease the asset from the government

BBO, buy-build-operate; BDO, build-develop-operate; BLOT, build-lease-operate-transfer; BOO, build-own-operate; BOOT, build-own-operate-transfer; BOT, build-operate-transfer; BROT, build-rent-own-transfer; BTO, build-transfer-operate; DBFO, design-build-finance-operate; DCMF, design-construct-manage-finance; LDO, lease-develop-operate; WAA, wrap-around addition.

Source: International Monetary Fund (2004).

The result of assessing the acceptability of PPP as a means of developing eco-tourist attractions by the government (staff of the ministries used for the study) showed that PPP as an option for the management of eco-tourist attractions in the state was accepted by 96.2% and while 3.8% want the private sector to manage these ecotourism attractions alone without government input. This implies that some of the staff that chose private management may be skeptical about their job security if PPP is used. All the staff who supported PPP believed that ecotourism attractions in the state are long overdue for development and that developing them will make them to be useful to the system and contribute their knowledge to tourism development in the state. They attested that government meager resources and lack of qualified tourism personnel can be curtailed through

PPP. This assertion is in tandem with the views of Dredge and Jenkins (2007) concerning tourism planning and policy. The result tallies with the findings from all the communities, where they voted 100% for PPP. They are of the view that since government has not shown much effort toward the development of these attractions, PPP offers them a lot of opportunity to benefit from what nature has bequeathed to their community. Furthermore, most of the informants revealed that in other states or countries where these eco-tourist resources are located the communities benefit a lot and it gives them a sense of pride. One of the informants in Agulu argued that if the lake is developed, that it will be better than developed lakes in southeast due to the crocodiles therein, and waiting for government is a mirage.

Similar studies across the globe have advocated for PPP as a viable tool for development of tourist attractions. It can be applied in management of forest reserves (Ngoka et al., 2014) and caves (Onwudufor and Odum, 2015). Tanzina et al. (2013) have advocated for PPP with the belief that it can augment limited government budgets. On a broader scale, Ekpenyong and Mmom (2015) have advocated for PPP and tourism in Niger Delta region of Nigeria. This makes it imperative for a state like Anambra, which has other developmental projects competing for the state meager or limited resources, to explore PPP as an option in developing its natural tourist attractions.

Outright privatization might make tourism services more expensive to the consumer and tourism development in the state seems to be at elementary stage, which means that most of the consumer are probably domestic tourists with limited income. Moreover, Anambra is surrounded by other states that have similar attractions that might be competing for tourist's attention. Therefore, outright privatization might shoot the price of privately eco-tourists destinations in the state high. This may affect the tourism market of Anambra.

The undeveloped nature of eco-tourist attractions in the state seems to scare any government that comes into power in the state, because of the 4-year tenure, every government is usually interested in executing projects in shortest possible time. The capital intensiveness of developing tourism, and the fact that its utmost benefits may yield in near future time and makes developing tourist attractions a "less-traveled" road by successive government.

Based on the aforewritten, it is pertinent to look at what it will take for PPP to thrive in Anambra state. PPP cannot succeed without appropriate regulatory framework, political commitment, and good governance. This will be followed by a working and efficient legal system in the state to assure private investors that agreement with government will be honored. Presently, government of Anambra state has shown elements of commitment in the development of tourism in Anambra state. For example, the government has tarred the road leading to and has put up some structures in Owerre-Ezukala cave and water fall in the state. Although this is a right step in the right direction, it is not enough. Currently, the government contracted hotel enumeration job in the state out to a private organization; a private firm (Golden Tulip) is managing the hotel built by government near Agulu Lake.

Figure 7.4 shows the first responsibility of government is to show willingness in developing tourism, and this is known by making tourism a priority, creating policies that will encourage private investors to participate in developing tourist attractions. Such policies must cover the host community interest to avoid conflicts. The government should provide tax rebate for tourism investors; this will encourage other investors. Government policies should ensure that investors are assured that their money is guaranteed in case government defaults in any agreement. Figure 7.3 shows different aspects of likely government responsibilities in Anambra State.

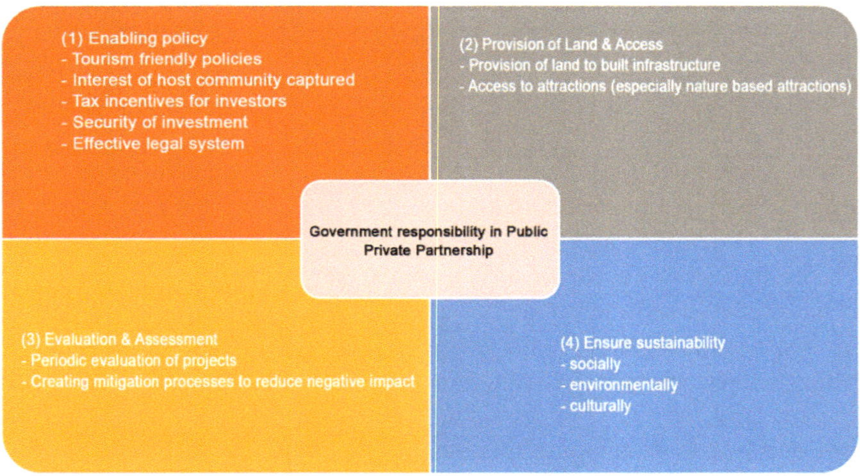

FIGURE 7.4 Government role in public–private partnership in Anambra state.
Source: Authors' creation.

This is followed by provision of land for building infrastructure. Government is usually the sole owner of land in Nigeria (see Land Use Act Decree No. 6 of 1978). Therefore, the need for government to give land to investors or lease is necessary here. These portions of land are where support facilities will be built. Compensation of communities will be the responsibilities of either government or the investors, depending on what the agreement between both parties says. Almost all the natural attractions in the state have no built accommodation facilities near them, except the recent hotel built close to Agulu Lake. Furthermore, government should create access routes to tourist attractions so that investors can come in. There are cases where the communities do not allow government and tourism stakeholders to come and access these attractions. Currently, kingship tussle has affected the development of Owerrezukalla cave and waterfall. Government should sue for peace in such communities before bringing investors.

7.8 EVALUATION AND ASSESSMENT

Private investors are usually motivated by profit unlike government. The quest for profit might make private investors not to adhere to laid down rules and regulations in developing tourism in the state; therefore, it is the responsibility of government to go for periodic evaluation and assessment of management standards in destinations that are given to private investors. Here, also, the interest of government workers should not be undermined by the private investors. The government should create ways or methods to ensure that negative impact of developing tourist attractions like environmental pollution, noise pollution, overcrowding, and so on are controlled and mitigated.

7.8.1 SUSTAINABILITY PRACTICE

PPP must ensure that developing eco-tourist attractions is in tandem with socio-cultural milieu in Anambra State, that is, environmentally compatible; the development is not polluting the ecosystem and is economically beneficial to all relevant stakeholders. For instance, the people of Agulu will not accept any development of the lake that undermines the cultural uses/values of crocodiles in their lake.

The private investors or sector in the PPP agreement should provide infrastructure. Majority of the hotels in Anambra state are private owned; there are no single accommodation facilities in close proximity to where these attractions are located, with exception of Agulu. There is no public convenience to cater for visitors and tourists where these attractions are located, except in Ogbunike cave and recently, in Owerre-Ezukala cave and waterfall. Roads leading to these attractions are untarred. All these can be provided by the private sector. There is shortage of personnel and lack of professionalism in public service. Therefore, the private sector should provide the necessary manpower and finesse to manage eco-tourist destinations in the state, sponsor trainings for staff and tourism stakeholders in the state for quality and efficient service delivery. Other concerns of the private sector include safeguarding of tourist attractions against misuse by community members and tourists, security of tourists, and visitors are essential to protect the image of the destination.

Furthermore, creating of database for tourist attractions in the state will help the private sector to know what type of attraction they are interested in and the locations. What is the profile of tourists/visitors coming to the state? Such database helps in planning and will give the private sector a clue of the profit margin they are expecting. The private sector should manage the advertisement of tourist attractions in the state as shown in Figure 7.5.

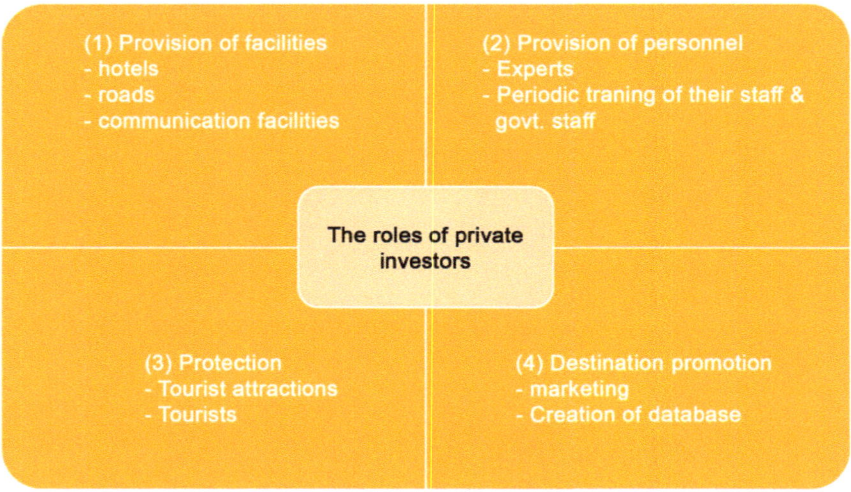

FIGURE 7.5 The role of private investors in PPP in Anambra state.
Source: Authors' creation.

7.9 CONCLUSION AND SUGGESTIONS

This study submits that PPP can be used to develop ecotourism resources in Anambra State given the prevailing challenges affecting tourism development in the state. Moreover, PPP has been used in developing tourist attractions in different parts of the world as mention earlier. Although, we support the use of PPP, its application should be adapted to suit the socio-setting of the study area. Given the present government poor attention to tourism and ecotourism resources in the state, seeking alternative means of developing these resources becomes inevitable. Second, community members and staff of related ministries support PPP as an alternative means of developing eco-tourist attractions in the state. Moreover, such partnership should ensure the interest of host communities are considered from the preliminary stage of such partnership to ensure sustainability. Furthermore, the built-operate-own option of PPP was equally advocated due the political terrain of the state. Conversely, we advocate the use of PPP for ecotourism development in developing nations such as Nigeria where government interest in tourism appears inconsistent.

7.9.1 RECOMMENDATIONS

The study recommends a further study should be carried out in line with exploring other variants of PPP other than the one suggested in this study as an option, and the demerits and merits associated with each. This can be followed by constraints associated with government and private partnership in developing tourist attraction. Such studies will aid this baseline we have created, and will be beneficial to policymakers and related tourism stakeholders.

7.9.2 SUGGESTIONS

We suggest that the Nigeria Conservation Foundation, National Park Services, Conservation International, IUCN, and so on need to be attracted to the state for further studies on these attractions.

KEYWORDS

- **Anambra**
- **attraction**
- **public–private partnership**
- **sustainable development**
- **tourism**

REFERENCES

Ajake, A.; Bisong, F. E.; Eze, B. E. A Preliminary Assessment of an "Uncharted" Ecotourism Asset: The Case of Esori Waterfall in Cross River, South Eastern Nigeria. *Am. J. Tour. Res.* **2013,** *2* (1), 95–100.

Aziekwe, N. E. Prospects and Constraints of Agulu Lake in Tourism Development in Anambra State. M.A. Thesis Submitted to University of Nigeria, Nsukka, 2008.

Dredge, D.; Jenkins, J. M. *Tourism Planning and Policy*; John Wiley: Milton, 2007.

Ekpenyong, O.; Mmom, P. C. Public Private Partnership and Tourism Development in the Niger Delta Region. *J. Environ. Earth Sci.* **2015,** *5* (2), 80–86.

Ewelum, I. C. Evaluation of Ecotourism Attractions in Enugu State. M.Sc. Project Submitted to University of Calabar, Cross River State, 2015.

Ezreth. PPP & CPPP Models for Sustainable Tourism Development in Kerala Tourism in Kerala: An Introduction. *Int. J. Bus. Admin. Res. Rev.* **2014,** *12* (3), 11–17.

Goloran, T.; Beeton, R. J. S.; Scott, N. *Grassroots Management of a Protected: The Puerto Princesa Subterranean River National Park, Philippines*; Global Report on Public-Private Partnerships: Tourism Development: Affiliate Members Report; UNWTO and Griffith Institute for Tourism (GIFT) 11, 2015; pp 20–22.

Hall, C. M. Rethinking Collaboration and Partnership: A Public Policy Perspective. *J. Sustain. Tour.* **1999,** *7*, 274–289.

Hall, C. M. *Tourism Planning Processes and Relationships*, 2nd ed.; Prentice-Hall: Harlow, 2008.

Herpen, G. W. E. B. *Public Private Partnership, the Advantages and Disadvantages Examined*; AVV Transport Research Center: Netherlands, 2002.

Ijeomah, H. M.; Alao, J. S. Assessment of Ecotourism Development at Farin Ruwa Waterfall in Nasarawa State, Nigeria. *J. Prod. Agric. Technol.* **2007,** *2* (3), 162–174.

International Monetary Fund. *Public-Private Partnerships*; Fiscal Affairs Department, World Bank and the Inter-American Development Bank, 2004. www.imf.org/external/np/fad/2004/pifp/eng (accessed Nov. 13, 2017).

Itanyi, E. I.; Okonkwo, E. E.; Eyisi A. P. A Preliminary Study of Cave, Rock-Shelters and Waterfalls in Owerre-ezukala, Orumba South Local Government Area of Anambra State, Nigeria. *J. Tour. Heritage Stud.* **2013,** *2* (1), 32–43.

Kumar, P.; Tiwari, P.; Mishra, J. M. Role of Public Private Partnership in Tourism Sector of Indian Economy. *Excel Int. J. Multidiscipl. Manage. Stud.* **2015,** *5* (8), 30–40.

Maria, J. Z.; Hall, C. M. Public-Private Collaboration in the Tourism Sector: Balancing Legitimacy and Effectiveness in Local Tourism Partnership: The Spanish Case. *J. Pol. Res. Tour., Leisure Events* **2012,** *4* (1), 61–83.

Mbaiwa, J., Ed. Stable Partnership Attract Funding, Botswana. In *Global Report on Public-Private Partnerships: Tourism Development*; Griffith Institute for Tourism (GIFT), Affiliate Members Report, UNWTO 11, 2015; pp 22–23.

Menon, S.; Edward, M. Public Private Partnership in Tourism: A Case Study of Kerala Travel Mart. *Afr. J. Hosp. Leisure* **2014,** *3* (2), 1–7.

Noga, C. K.; Geoffrey W. Evaluating Tourism Potential: A SWOT Analysis of the Western Negev, Israel. *Tour. Rev.* **2007,** *55* (1), 51–63.

Odum, C. J. Community Involvement in Tourism Planning and Development: A Study of Selected Tourism Attractions in Anambra State. Unpublished M.A. Thesis Submitted in University of Nigeria, Nsukka, 2011.

Odum, C. J. Ecotourism Survey of Awka-South. Ph.D. Seminar Submitted in Department of Geography and Environmental Science, University of Calabar, Calabar, 2014.

Odum, C. J. Evaluation of Potential Ecotourism Attractions for Sustainable Tourism Development in Anambra State, Nigeria. Ph.D. Dissertation Submitted in University of Calabar, Cross River State, 2017.

Odum, C. J. Tourism in Anambra: Today and Tomorrow. A Workshop for Tourism Department, Ministry of Diaspora Affairs, Culture and Tourism, Anambra State, 2018. Available from: www.ottourism.blogspot.com.ng [accessed March 24, 2018].

Odunlami, S. S. *An Assessment of the Ecotourism Potential of Yankari National Park Nigeria.* Eco-club E-paper Series, 2003. http://ecoclub.com/library (accessed Jan. 28, 2015).

Oguamanam, C. C.; Nwankwo, E. A. Sustainable Development Plans for Caves in Southeast Nigeria for Tourism. *J. Tour. Hosp. Sports* **2015,** *3*, 1–6.

Okeke, I. O. C.; Nwokolo, O. C. In *Conservation and Preservation of Traditional Agulu Lake in Anambra State, Nigeria*, Proceedings of Taal 2007: The 12th World Lake Conference; Sengupta, M., Dalwari, R., Eds., 2008; pp 2209–2211.

Okonkwo. E.E.; Odum, C.J. Tourism: An alternative for Economic Independence and Rural Development in Nigeria. *Africa Journal of Pure and Applied Science* **2010**, *5*(3), 48–55.

Onwudufor, F. O. F.; Odum C. J. An Ethnographic Excursion to Orumba: A Preliminary Report on Ufuma Cave and its Tourism Potentialities. *J. Hum. Soc. Sci.* **2015,** *20* (II), 13–20.

Peter, F. Working Together: Assessing Public-Private Partnership in Africa. In *Nepad Policy Focus Series*; The South African Institute of International Affairs: Pretoria, 2005.

Smith, A. L., Ed. In *Public-Private Partnerships (PPP) for Sustainable Tourism*, Interamerica Travel Congress, San Salvador, September 29th–30th, 2011. Inter-American Development Bank, Organization of American State, 2011.

Tanzina, C.; Syeda, T. F.; Evana, N. D. Promoting Public Private Partnership for Development of Tourism Sector of Bangladesh: An Exploratory Study. *Eur. J. Bus. Manage.* **2013,** *5* (32), 194–202.

Ukpaka, C. G. A Case Study on Current Status of Some Plant Species of Anambra State. *J. Occup. Saf. Environ. Health* **2014,** *2*, 234–238.

CHAPTER 8

TALENT MANAGEMENT CHALLENGES FOR HERITAGE TOURISM AND RURAL SUSTAINABILITY: A CASE STUDY OF SURAJGARH FORT HERITAGE HOTEL, INDIA

EKTA SHARMA[1] and SANDEEP SHARMA[2*]

[1]*Ahmedabad University, Navrangpura, Ahmedabad, India*

[2]*Magnum Opus International, Thaltej, Ahmedabad, India*

[*]*Corresponding author. E-mail: sandyshar15@gmail.com*

ABSTRACT

Thakur Suraj Mal constructed the fort of Surajgarh in 1778. The Fort is now owned by Tikaraj Aishwarya Katoch and Tikarani Shailija. Tikaraj Aishwarya is the son of Princess of Jodhpur and Maharaja of Kangra. They have converted the fort into the hotel. Although the fort is in the Shekhawati Region of Rajasthan, which is a great tourism site because of splendid havelies and frescoes but Surajgarh is a small village and it is a great challenge to attract, recruit and retain the employees for the Surajgarh Fort Hotel. In this chapter, the author will analyze and evaluate the talent management challenges in "rural heritage hotels."

8.1 INTRODUCTION

India has always attracted foreign tourists due to its rich cultural and built heritage. In particular, rural India has very large heritage assets. The growth of the tourism industry has opened up avenues for the development of "rural tourism" in India. Proper planning and management of tourism can be a tool for rural sustainability by supporting the socioeconomic development of localities as it will lead to employment generation (Varma, 2015). International tourist arrivals increased by 7.0% in 2017, the highest growth since the 2009 global economic crisis and UNWTO's longest forecast of 3.8% per year for the period from 2010 to 2020 (Highlights of UNWTO Tourism, 2018 edition). Tourism has become an important sector in the world economy contributing around 9% to the world's GDP. India had 10.04 billion foreign tourists with a 14% annual increase in 2017 (ITS Nazar, 2018). For the development of rural India, various stakeholders must come together to strengthen tourism.

Rural tourism facilitates regional development through creation of jobs and income for the local people, consumption of local products by the tourism sector, support to local handicrafts and other economic activities, and improved infrastructure and facilities.

One of the major tourist attractions of India is the Shekhawati region of Rajasthan, which is known for its magnificent "havelis" and "fresco." The old forts and havelis have been converted into "hotels" and are called "heritage hotels." For the last four decades, some architecturally distinctive properties such as palaces and forts built before 1950 have been converted into hotels that have been classified as "heritage hotels" by the Ministry of Tourism, Government of India. About 32% of classified heritage hotels are concentrated in Rajasthan (20%) followed by Kerala. The largest number of heritage hotels is located in Rajasthan (70) (Tourism.gov.in/sites/default/files/Other/Heritage% 20HOTEL.pdf).

Although these heritage hotels attract tourists both national and international, the major challenge for these hotels is to attract the right talent to manage the property. Managerial positions are difficult to fill due to nonattraction for employment in rural areas, mainly due to lack of basic amenities and other facilities. Urban youth with a degree in hotel management will not want to settle in rural India because of these challenges. Most of the time, even though managerial positions are filled, the retention rate is very low because whenever the manager gets better opportunities in the urban area, he leaves the job. In addition, they prefer to take their families to the rural site where they are working, rather than settling in nearby cities. Most of the time, this is due to the better educational, medical, and employment facilities available in these cities. This leads to high absenteeism among employees which is another challenge.

8.2 HUMAN RESOURCE AND HERITAGE MANAGEMENT

The human resource function varies across different industry and sectors (Poór et al., 2009). The divergent (contextual) perspective accepts that national, cultural, and sectoral differences affects the management techniques and also the human resource management (Hofstede, 1980; Ronen and Shenkar, 1985; Due et al., 1991; Daley, 2002; Brewster, 2004; Morley, 2004).

The heritage sector employee requires fresh attitudes, specialized training and experience, and distinct competence. Hence, the HR policies and practices for the heritage sector needs to have a different perspective. Right from acquiring, developing, evaluating, and compensating, all practices have to be framed to fulfill the requirement of the heritage industry. The Job and roles for the employees need to be redefined. For example, a role and job of room-service employee for the hotel industry in general and the heritage hotel in particular would have vast differences. Although the broad structure would remain same but the specific competence to suit the heritage requirement need to be customized. The essence of hospitality would remain same but the mannerism would change and hence, recruitment, selection, development, and rewards need to be redefined to align with the organization requirement. "Our understanding of skills encompasses a bundling that goes beyond the technical to incorporate emotional, aesthetic and informational processing and analysis dimensions with a strong focus on the delivery of service to diverse consumers" (Baum et al., 2007).

8.3 TALENT MANAGEMENT

Talent management is the human capital which contributes to the organizational success (Dries, 2013) and the talent lifecycle is a representation of the stages of interaction between the organization and its human capital, which begins with attracting the right individuals passing through the stages of "...acquiring, onboarding, developing, managing, retaining and even recovering talent" (Schiemann, 2014).

Talent management seeks to attract, identify, develop, engage, retain and deploy individuals who are considered particularly valuable to an organization. By managing talent strategically, organizations can build a high performance workplace, encourage a learning organization, add value to their branding agenda, and contribute to diversity management. For these reasons, HR professionals widely consider talent management to be among their key priorities. (CIPD, 2017)

Talent management is important for the organizations as the "War for talent" is immense. The talent pool is considered the competitive advantage and to keep it intact, organizations need to have employee engagement and retention policies in place. The organizational culture also facilitates the attraction and retention of the top talent. The replacement and the succession plans need to be of utmost importance to be able to have continuous coverage of critical roles.

Talent management would broadly include talent planning (understanding the organizational/business strategy, developing Measurement matrix, workforce planning); talent acquisition (employee-value proposition, employer branding, talent acquisition); talent development (orientation or on-boarding; performance management; learning and development; potential appraisals or capability frameworks; career planning); talent retention (organizational culture, compensation and reward strategy); talent transit (succession planning, transfers, retirement, knowledge management, exit interviews).

8.4 FACTORS INFLUENCING TALENT MANAGEMENT DECISION

8.4.1 TALENT SHORTAGES

The unemployment rates are high but the skilled workforce is in shortage. There is need to increase the talent pool.

8.4.2 DEMOGRAPHICS AND SOCIETAL TRENDS

Research has highlighted rapid shifts in the demographic profiles of many countries which impact on the supply of labor available to employers in those countries, and this requires many organizations to introduce new approaches in their recruitment and retention policies and practices (Tarique and Schuler, 2010).

8.4.3 THE INCREASING MOBILITY

The increasing mobility of people across geographical and cultural boundaries progressively impacts on decision making in global talent management (Tung and Lazarova, 2007; Vance et al., 2009).

8.4.4 THE WORK

The workplace is characterized by creating new forms of work as well as new jobs, by the disproportion between the required and available skills, by the shortage of high-skilled workers, and by the rise in the global and virtual workforce. Technology has made the world faster and more connected, has a significant effect on workplaces and the workforce.

8.5 CASE STUDY: SURAJGARH FORT—HERITAGE HOTEL

Surajgarh is a town in Surajgarh Mandal, Jhunjhunu District of Rajasthan. Surajgarh is 37.5 km far from its district's main City Jhunjhunu, 154 km far from Jaipur and 180 km away from Delhi. Surajgarh is located at 28.32°N 75.73°E. As of 2001 Asian nation census, Surajgarh had a population of 18,857. Surajgarh is famous for the frescoes, which depict different stories. They owe the painting of frescoes to the travelers who came through this region, bringing along with them stories of the lands far away which got its expression by the local populace who started to decorate their homes with frescoes of the tales told by the travelers and also from the mythological scenes. The materials used for painting the frescoes were crushed cowry shells, lime, and vegetable dyes which were produced locally. The cowry shells were used as currency at only once, but since the chipped and broken shells had no value, they were crushed to produce a water-resistant base for the frescoes.

Surajgarh is in the center of Shekhawati's open-air art gallery and has many havelis. Marwari traders accumulated monumental wealth in their trade of rice, opium, cotton, spices, and textiles. Money was sometimes remade or sent back home for himself or his family. The thriving Marwari families attempted to do the exact opposite of the size and quality of their mansion, the division for fine and lively design within the desert region wherever least expected. Wherever the painting was done, the lime plaster on the walls shaped a fine layer. Although the plaster was moist, outline, and color work was done, it was added for the premature. Tourist walk and camel cart tour are organized for havelis.

In 1778, Thakur Suraj Mal built the fort of Surajgarh. Presently, the fort is closely guarded by Tikraj Aishwarya Katoch and Tikrani Shalia. Tikraj Aishwarya is the son of the aristocrat of Jodhpur and the prince of Kangra. They have converted the fort into a hotel. However, this fort is within the Shekhawati region of Rajasthan, which is a great tourist destination due to the magnificent havelis and frescoes. Surajgarh is a small village, and attracting, recruiting, and retaining staff for Surajgarh Fort is a good challenge.

The Thakurs of the East sold Surajgarh Fort just after independence. It has since changed many hands. The Kangra group of hotels bought it from the developers of the land who were trapped with the locals and the collector stopped them from demolishing the palace. Later in 2018, it was sold to the "royal family of Patiala."

8.5.1 EMPLOYEE POLICIES

Since it is in a remote area, special attention needs to be paid to employee selection and retention. They appoint trained general managers and departmental heads who then hire and train local people.

Surajgarh is situated in rural areas, where professionals have less opportunities and facilities and will neither like to work long nor settle their families. The management accepts that this is a challenge and goes on to say, "Yes, we do have a problem with hiring well-trained staff, but as Rajasthan is a tourism-driven industry, there are many small hotel management institutes that own Serve Surajgarh with staffing requirements. GM usually comes without his family. As we tie up with marketing and management consultants with a successful record of partnering with some of the leading travel and hospitality brands in the Indian subcontinent, most management-level positions are filled by these consultants. Housing and food are also provided to employees." But the retention rate is low.

8.5.2 DISCUSSION

The rural tourism and sustainability require an integrated planning and development approach which takes into consideration all the factors necessary for the long-term growth and development of the region, especially the human capital capacity building and infrastructural developments. Capacity-building activities would develop the knowledge and skills of the rural people enabling them to make a career in the tourism sector. Proper and planned investment in infrastructural development would facilitate the tourist, talent attraction, and the locals.

The problem of talent attraction and retention is evident at "Surajgarh Fort Hotel." Although the recruitment and selection process has been outsourced still it is difficult to hire the talent especially at the management positions. One of the reasons for the turnover at management positions is that the GMs do not find the infrastructure appropriate as per their living standards and basic requirements. Since the hotel is located at the remote site of Rajasthan, it lacks sophistication and amenities. If the employee is not the local person, he/she would not prefer to locate his/her family to this remote place, as the spouse and children would not be able to adjust to the rural setting and also the education and healthcare would be a matter of concern. With this kind of instability, the employee would explore the other opportunities where he/she could avail better amenities. The following talent management techniques could help the hotel to retain its employees: edition of "in between" experiences in cultural tourism, encompassing elements that do not belong either in the local nor in the tourist sphere (Richards, 2018).

At the same time, communities have been increasingly outspoken in their opposition to mass tourism, but the majority of destinations seem to have been unable to provide a comprehensive, scalable alternative. The recent insistence by New Zealand and Hawaii on attracting "better tourists," who would be willing to make real connections with locals—and generate more value to the economy—seem desperate and elitist at the core (Mathews, 2019). What seems to drive the tourist–local divide is a particular misunderstanding surrounding, at the street level, of the value of "the everyday." The antagonism is also often exacerbated by the relations of dependence created by international tourism (Nogués-Pedregal, 2019).

8.5.3 EMPLOYEE VALUE PROPOSITION AND EMPLOYER BRANDING

The first strategy should be to propagate the value proposition to the potential employees about working with Kangra Group and Surajgarh Hotel. The group is gradually becoming a brand in the tourism sector, thus the benefits of working with the hotel should be communicated clearly. This would help in attracting the talent pool.

8.5.4 ACQUISITION

Further, being a heritage hotel, the concept itself is "unique," yet known. Thus, the passion for working towards its betterment through unique ideas of maintaining it and bringing it out to the public is important. For the lower level administrative and management position, passion to work for heritage site is preferred over a high level of education. Therefore, even the locals of the town could be the most resourceful employee bringing the benefits like: (1) since they are locals they would be aware of the context, (2) they would have a rich thread of history, knowledge, and situations of the place as well as the surrounding, (3) they would get a better source of livelihood and an assurance of work in their own town, which would help in retaining the staff.

For the middle management positions and other professional positions like "Head Chef," they could tie-up with the hotel management institutions in the vicinity of the town. Also, it would be advisable to have tie-ups with some good educational and healthcare institutions like "Birla School" and "Birla Hospital" at Pilani, which is less than half an hour's drive from Surajgarh. Even the spouse could be engaged in some activities of the hotel or some corporate social responsibility functions like teaching the rural children or educating about hygiene, which would not only engage the spouse but also give the "sense of accomplishment" and "feeling of respect" for the hotel.

8.5.4.1 CAREER PATHS

Job retention translates into profits only if the career path is clear for the employees. The options to move up the ladder and the requirements of every level should be communicated to the employee and should be allowed to pursue their career path.

8.5.5 PROVIDE FACILITIES

The very first concern for the candidate applying for the job in a small town is a concern for the partner or family. They will have to live far from their family. The family cannot be brought here because of the lack of facilities like school, hospital, banks, malls, and all. If such facilities are provided the employee can be attracted and retained for a longer time period. If facilities in the near surroundings are facilitated, the concern to live in small town will reduce.

8.5.6 SALARY AND BENEFIT PROGRAMS

Working in small cities is a concern nowadays for employees leaving the city lifestyle and amenities. To attract them with higher salary from other organizations of same sector located in city and to provide additional incentive offers along with the salary, the organization needs to offer employees at least life and medical coverage. The company can collaborate with insurance companies for employee betterment so that they offer more inexpensive insurance. Further to give employees a stake in a company's success, to increase the sense of ownership, profit sharing in form of employee stock option plans could be offered. Partnering with local businesses to offer the team members discounts in the areas of healthcare, professional services, education, home ownership, retail, and entertainment, the hotel goal should be to cultivate an environment of respect and dignity for everyone who chooses to join the hotel's team.

8.6 ROLE OF THE MANAGEMENT IN ACHIEVING SOCIAL AND CULTURAL SUSTAINABILITY

8.6.1 REDEVELOPING THE TOWN

The aim to recruit the local community members for service staff can be done by developing the environment and other facilities in the town. The objective is to use the nonutilized land for development. The standard of living of the community and quality of life would increase and would become better. Providing basic facilities like school, bank, and hospital would increase their standard of living. Community involvement is a key to revitalizing the town.

8.6.2 DEVELOPING THE LOCAL PEOPLE

The local people could be trained to take up employment in hotel, which would support the financial viability of the local houses.

8.6.3 OPPORTUNITIES TO THE LOCAL ARTISANS AND CRAFTSMEN

The local artisans and craftsmen could be given opportunity to showcase their art, thus would not only provide employment opportunities to the artisans, but also would help in cultural sustainability.

8.7 COLLABORATION

Collaborations with other organizations dealing with heritage can be done to organize cultural programs and events. Income can be generated through these events. Organizing events during heritage week and programs like music festivals where the local art and artisans can be promoted along with the fort. The local art and craft can also be promoted and financial aid can be generated by developing a shop for local craft and art and involving the local community as well. It will also promote and protect the traditional local community craft and art. For example, at the Mehrangarh Fort, a music festival is organized for three days at the fort where artist from various places come to perform and visitors from around the world come to see the festival. Apart from regular visiting charges, this is the only major source of income for the fort and organization where they sustain for the whole year. The Corporate events like "ICOMOS General meeting about Heritage and Tourism in Surajgarh Fort" or Yearly conference tie-ups with Institutes could also be the source of income.

8.7.1 MUSEUM OF PERSONAL FAMILY COLLECTIONS

There is a huge potential to turn some of the unused portions of the fort to a museum that could be created to tell the fort stories through the objects and memories associated with the fort and the town. Museum shop can sell the arts and crafts of the local artisans such as reprints of frescos, lac bangles, bandhani and block printed textiles, and other handicraft items of the region.

8.7.2 EXHIBITION AND ART GALLERY

Hotels can let out the rooms to local entrepreneurs, artists, and fashion designers for various exhibitions on rent during the peak tourist seasons.

8.7.3 CENTER FOR DESTINATION EVENTS

Although the tourist activities are at its low, the hotel offers ideal location for film shoots, fashion and other advertisement shoots, events, and weddings, engagements, anniversaries, and school or college reunions. This can be highlighted on many marketing portals. These can be made into attractive packages and be a win–win situation for all, the visitors, the business people, and Surajgarh Hotel.

8.7.4 HERITAGE WALK

Shekhawati havelis are well known for frescoes and Surajgarh also has beautiful 200 to 300 years old frescoes. These can offer a golden opportunity for the hotel to plan a heritage walk around the hotel and town to give a complete collage view to the hotel and town to its visitors. There are other old buildings and wells which could be included in the heritage walk.

8.7.5 LOCAL ARTISANS, MUSICIANS, PERFORMERS, STORY TELLERS

They can be encouraged to come and perform at the hotel for the guests and this activity can be an ongoing activity with everyone participating in it. This will generate employment opportunity for the village communities and create better economic opportunities for the hotel as well. There could be live craft shows where the artisans explain the audience how they perform their art. Art of pottery, making ornaments of lac, and so on could be shown to the tourists.

8.8 CONCLUSION

The fort palaces are now converted into heritage hotels and majority of these hotels are located in small towns. To run and maintain these forts is a big task for the management and trustees both. The rural sustainability should be the key here. The efforts to develop, support, restore, and revive the cultural as well as the built heritage of the town should be the focus. It is difficult to get employees for managerial positions in these types of organizations because of the situation of the organization and surrounding. Still many small towns in the country are under developed where still the basic facilities are not available to the local community. Government and private sector both should collaborate to overcome these challenges and to develop those small towns and the hotels worldwide. The idea to promote the heritage of India and its culture should be the strategy to sustain the heritage.

KEYWORDS

- **heritage hotel**
- **fort**
- **talent management**
- **employee policies**

REFERENCES

Baum, T.; Dutton, E.; Karimi, S.; Kokkranikal, J.; Devine, F.; Hearns, N. Cultural Diversity in Hospitality Work. *Cross Cult. Manage.: Int. J.* **2007,** *14* (3), 229–239.

Brewster, C. European Perspectives on Human Resource Management. *Hum. Res. Manage. Rev.* **2004,** *4*, 365–382.
CIPD. Talent Management. *Understanding the Changing Context and Business Case for Talent Management and the Key Features of a Talent Management Strategy*; 2017.
Daley, M. *Strategic Human Resource Management*; Prentice Hall: Upper Saddle River, 2002.
Dries, N. Applying a Talent Management Lens to Career Management: The Role of Human Capital Composition and Continuity. *Int. J. Human Res. Manage.* **2013,** *24* (9), 1816–1831.
Due, J.; Madsen, J. S.; Jansen, C. S. The Social Dimension: Convergence or Diversification of IR in the Single European Market. *Indus. Relat. J.* **1991,** *2*, 8.
Hofstede, G. *Culture's Consequences: International Differences in Work Related Values*; Saga: Beverly Hills, 1980.
ITS Nazar. 2018. http://tourism.gov.in/sites/default/files/Other/ITS_Glance_2018_Eng_Version_for_Mail.pdf (retrieved on June 14, 2018).
Morley, M. J.; Collings, D. G. Contemporary Debates and New Directions in HRM in MNCs. *Int. J. Manpower* **2004,** *6*, 487.
Poór, J.; Karoliny, Z.; Bártfai, B.; Pótó, Z.; Farkas, F. Similarities and Differences of Human Resource Management in Private and Public Sector Organisations in the Light of New Public Management in International Comparison. *Acta Oecon.* **2009,** *59* (2), 179–206.
Ronen, S.; Shenkar, O. Clustering Countries on an Attitudinal Dimension: A Review and Synthesis. *Acad. Manage. J.* **1985,** *10* (3), 435–454.
Schiemann, W. A. From Talent Management to Talent Optimization. *J. World Bus.* **2014,** *49* (2), 281–288.
Tarique, I.; Schuler, R. S. Global Talent Management: Literature Review, Integrative Framework, and Suggestions for Further Research. *J. World Bus.* **2010,** *45* (2), 122–133.
Tung, R.; Lazarova, M. The Human Resource Challenge to Outward Foreign Direct Investment Aspirations from Emerging Countries: The Case of China. *Int. J. Human Res. Manage.* **2007,** *18* (5), 868–889.
Vance, C. M.; Vaiman, V.; Andersen, T. The Vital Liaison Role of Host Country Nationals in MNC Knowledge Management. *Human Res. Manage.* **2009,** *48* (4), 649–659.
Varma, R. R.. Heritage and Development. *Rural India* **2015,** *3* (1), 15–23.
UNWTO Tourism Highlights, 2018 Edition.
www.kangragroup.com.
www.surajgarh.com/eventsinsurajgarhfort.shtml.

CHAPTER 9

TRIBAL TOURISM DEVELOPMENT THROUGH TRIBAL FAIRS IN RAJASTHAN, INDIA

SURABHI SRIVASTAVA

Associate Professor, Om Kothari Institute of Management, Kota, Rajasthan, India. E-mail: surabhi3k@gmail.com

ABSTRACT

Tribal fairs are the mirror of the cultural and historical bequest of the community. Geographically and socially isolated, these fairs are linked with visualizing tribal cultural diversities. Their customs, traditions, attire, music, and dance are the keys to their social and economic development where their originality and authenticity are preserved reverentially. In reality, these fairs are also positively contributing to the country's development process. Thus, the objective of this research is to study the impact of tribal festivals on the local and wider socioeconomic well-being. The endeavor is to find out the role of tribal fairs in their social and economic development and to dig out the possibility of involvement with the mainstream of the society. Rajasthan is a state of India that reflects various colors of cultures and traditions. It also comprised many tribes living with their originality and authenticity. This research takes case analyses of tribal fairs of Rajasthan. Baneshwar fair by the Bhil tribes and the Sitabari Mela by Saharia tribes are dominant in this area and our target group. This literature-based research affirms that each tribe has its traditions and customs. Tribal fairs are the showcase of their complete culture and cultural tourists always admire the originality. So it can be a tool for their social and economic development. Many strategies need to create an environment for tribal fairs capitalizing opportunities. Finally, this chapter suggests the need to preserve the intangible tribal heritage through tribal fairs.

9.1 INTRODUCTION

Tribal tourism is a new and expanded addendum, which is in such postmodern travel mode as the journey of postmodernism, nature, and wilderness. For tourists, these new destination concepts present a vague awareness of dwindling resources that individuals should see while they still can (Smith, 1996). "Tribal tourism" has been defined by Ward (2010) as "a new form of travel," in which tourists travel to tribal villages to be completely isolated from their culture. India, Madhya Pradesh, Jharkhand, Orissa, Nagaland, Himachal, and Chhattisgarh are states that can develop more and more tribal tourism. Tribal tourism has been instrumental in creating many money opportunities for tribes living in mountainous areas. It has helped to raise awareness about indigenous people in India, many of whom face harassment, lack of opportunities, and social exclusion. Although tourism may sound good to society, it is also causing social and psychological consequences for tribes that are more harmful than beneficial.

Many researchers, NGO activists, social workers are working for tribal development, upliftment, and inclusion in society. Providing basic amenities to the essential community is never the solution. Keeping this idea in mind, the tribal tourism concept came to the mind of many researchers. As the original tribe area unit wealthy with ethnic characteristics of cultural resources, some remote tribe in particular has preserved many of the social systems, handicrafts, and rituals that individuals within the city build a strong attraction. Chang et al. (2009) found that native culture is a very important indicator of tourism product mirrors based primarily on tribal tourism areas, which is basically the main material of the original goods, as it has become the replacement style of economic sources. Hinch and Manservant (1989) found that native tourists suggest that natives are directly related to the conduct of tourism trade or tourism trade as a way of attracting tourists to return to tribal culture. Research is mainly conducted from literature review and on-site visits to achieve the set objectives of the subject; that is, tribal festivals can be a tool for social and economic development and preservation of their customs and tradition. The purpose of this study is to find out whether tribal festivals can support their economic development on the one hand and the preservation of their customs and traditions on the other.

"Indigenous Business Analysis, Past and Present: Where to From Here?" written by Whitford and Ruhanan (2016) where the results show that sustainability issues shape a significant proportion of published and indigenous tourism research to date. The challenge may now be to gain an additional comprehensive understanding of native tourists from the angle of native stakeholders, approaching its complexity in an iterative, adaptive, and flexible style, and within the analysis methodology, knowledge creation, and its consequences for the affected stakeholders. It is both an ethical imperative and a pragmatic approach that facilitates the sustainability of indigenous tourism to ensure research results.

Carr et al. (2016) highlighted the potential of tourism as an effective tool for realizing sustainable Indigenous development in "Indigenous Peoples and Tourism: Challenges and Opportunities for Sustainable Tourism." Throughout the papers reviewed intensively, readers are reminded of the positive (capacity building) and negative (revision) realities of indigenous tourism development.

A section of the "Commodification and Management of Culture, Advance in Tourism Research, 2005, Tourist-Host Nexus-Research Things" in Ryan and Aicken (2005) examined the nexus of the researcher and conducted research in the context of the relationship between tourism and indigenous peoples. It was argued that three dimensions exist in this regard, such as education, indigenous perspective, and the imperative of tourism.

9.2 INDIGENOUS TOURISM

The article "Indigenous Tourism Development in the Arctic" by Claudia Notzke (1999) explores current trends in indigenous (tribal) tourism development in the Western Arctic region of Canada. In the north, tribal tourism is a resource-based industry, traditionally in the form of hunting for big game and, in a more modern context, evolves into ecological and cultural or ethnic tourism. Some endemic people search for new methods to support the standard components of their land-based economy rather than being consumed by class industry. The authenticity of this tourism expertise represents a major asset as well as a significant management confrontation.

"Attracting tourists to tribal cultural festivals: An example in Rukai tribal area, Taiwan" by Chang (2006) reported that festivals are increasingly being used to promote tourism and boost the regional economy. The area unit of festivals is a type of cultural event and the area unit is a travel attraction with specific options. A lot of research, done from different perspectives, exists on festivals. However, very few studies related to tribal cultures have been published. The main objective of this study is to profile tourists based on their motives and demographic characteristics, as these traits are class measurements related to attractions for tribal cultural festivals and other related activities. Report analysis showed that cultural motivation is, among other motivational dimensions, the most important factor attracting tourists to the tribal cultural festival. Furthermore, not all tourists are equally interested in competition cultural expertise. In addition, motivational variables are found to be more important than convincing visitors to a visitor festival and demographic variables in the segment.

9.3 TRIBE AND HISTORY

The English word tribe comes back from the Latin word "tribus" which denotes a specific type of general and political organization that has been conveyed to all societies. The name Individuals' tribe refers to a class of individuals and refers to a stage of development in human society. These tribes depended heavily on the forest for their daily needs, including food, shelter, equipment, medicine, and in some cases clothing. As long as the tribes were in control of the forest and its products were used for open use, they had no difficulty in meeting these needs. In return, he protected the forest because it was his life network. During the last 60 years and especially during the planning period, the pace of mining and manufacturing industries, as well as the use of power and forest resources, has been substantially accelerated, which is responsible for the loss of their indigenous culture and this is partially documented in the book "The Tribal Culture of India" (Vidyarthi and Rai, 1977). This study is shedding light on the important issues related to their cultural life that tribes face today. Although industrialization and modernization have their own importance, we also have to realize that tribal culture has to be preserved, buried in the greater focus on industrialization and modernization.

Special interest tourism (SIT), a rapidly growing concept and approach to tourism, demonstrates a strong people-centered, sustainable model that seeks a more authentic experience to enable closer contact with host communities. Weiler and Hall (1992) suggest that rewarding, enriching, thrilling, and learning are the four key elements of special interest tourism. SIT products offer customized packages of experiences developed for niche markets (Douglas et al., 2001). These products provide unique insights into particular sites, communities, or bodies of knowledge. Under the purview of SIT, indigenous people are now becoming part of the tourism industry. This section named "Indigenous Tourism" makes it a special interest area for tourists who like to experience authentic indigenous culture and ethnicity.

Indigenous tourism is a type of resource-based tourism type, and as a major tourist attraction, the development of tourism in addition to indigenous culture, local scenic natural landscapes and exotic landscapes, and a lot of tourists consider the "man" to be truly necessary. Many scholars have defined various definitions for tribal tourism: Ryan and Houghton (2002) define tribal tourism as "a tourist artistic display by tribal culture, festivities, attractions, historical heritage, and customs tourists engaged in tribal areas Attracts travel activities." Wu (2003) believes that "tribal tourism" differs from the general pattern of tourism activities, as contacts are unable to ride hard or talk with natural resources, but those who live, authentic cultural traditions. "Tribal tourism" is defined as the development of tribal cultural resources as a pivot, which are developed things; statements of activities include tribal arts and crafts, clothing, architecture, music, dance, cultural traditions, and so on (Liao and Lin, 2003, "A Study of Indigenous Tribes Tourism Developing-Case by Lilang, Tbulan, and Hrung in Taiwan," 7).

Therefore, autochronic business ventures typically involve small businesses that support the underlying social group information of culture and nature. It flourishes when indigenous people operate tourism and cultural centers, provides visitor facilities, and control cultural events and tourist access to homelands (Zeppel, 2001). According to Smith (1996), four Smith Hs (e.g., habitat, heritage, history, and handicrafts) define indigenous tourism as a segment of the visitor industry that directly includes native people whose ethnicity is a tourist attraction. Indigenous tourism is a means of cultural survival for many Native communities and provides a way to overcome social isolation.

Tribal cultures are of special interest and primary motivating factor for tourists traveling to foreign places, regions, attractions, and events. These include indigenous museums and cultural villages, nature-based tourism, indigenous festivals and programs, and indigenous art galleries. Indigenous tourism attractions are often located in rural or remote areas with limited infrastructure (Getz and Jamieson, 1997). Cultural, environmental, and spiritual aspects of indigenous heritage and traditions make it an integral part of indigenous tourism (Heather, 2006). The primary strength of tribal tourism lies in its ability to attract tourists and arouse their interest to experience local cultures in the most authentic settings. The participation of tourists in interactive activities with tribal people is another important strength of indigenous tourism. The priority of this market is more in offering intimate and spontaneous cultural experiences as there is a strong connection between tribal cultures and the natural environment. The Tribal Cluster inherits the associate degree of traditions, which are deeply frozen in their culture and lifestyle.

9.4 TRIBAL VARIATIONS IN RAJASTHAN

India is a land of festivals, cultural fairs, and overall seasonal congregations. These festivals help to strengthen cultural roots and values, enable communities to maintain their traditions, and bring economic values. All these auspicious festivals have their roots in economic agents such as agricultural enjoyment, business prosperity, entrepreneurial development, and social benefits. Rural fairs and festivals allow local communities to engage with their talent, arts, crafts, food, dance, and other forms of cultural landscapes and promote significant livelihoods through entrepreneurial activities. They bring about an economic boom through a culture of exchanging gifts, sweets, and fruits, which are entirely linked to economic agents such as producers, entrepreneurs, and consumers. In every festive season, the demand for specific events and services is driven by cultural events, and the supply moves according to demand and on a larger scale. Thousands of shopkeepers who live in a makeshift shop on the streets get new opportunities in festive times.

The people of Rajasthan are undoubtedly an interesting aspect of this part of the world. All communities differ in costumes, customs, occupation, and accommodation. These include a wide range of lifestyles, ranging from inhabited nomads, aristocrats to tribal shepherds, farmers, merchants, craftsmen, and camel-shepherds. Such communities stand for symmetry in esthetic and socioeconomic relations. Traditional tribal costumes are the hallmark of local people living in Rajasthan communities. With modernization, change is inevitable. Although pieces of jewelry, clothing, and adornments may be similar, individuals have developed distinctive costumes. They like to wear their traditional dress and show the world their true identity.

The Bhils are known as the Dhanush men of Rajasthan. They are the team of the greatest social group in the Asian nation. They like the most important tribe in the whole of South Asia. The Bhils live mainly in the western region of Madhya Pradesh as well as neighboring states of Gujarat, Rajasthan, and Maharashtra. Bhils are also found in the northeastern parts of Tripura.

9.4.1 BHIL

The Bhil tribe belongs to the caste of former Aryans. The name "The Bhil" is derived from the word Villu or Billu, which according to the Dravidian language is known as Bo. They are probably the oldest tribes of India and are described in Hindu mythology and also consider Valmiki and Eklavya as their ancestors. During the traditional era, they were thought of as good warriors as the United Nations (UN) agency fought against the Mughals, Marathas, and the British.

The language and costumes of the Bhil tribe are also attractive. Among other states, Bhils account for 39% of the total population of Rajasthan. Speak the Bhilli, which is an Indo-Aryan language. Bhil women wear ancient saris and hence Bhil men wear loose long frocks with pants. Farmers wear turbans. The Bhils also wear brass jewelry. He is known for his veracity and simplicity. He is brave and his national weapon is a bow made of bamboo. At first, they were great hunters. At present, they supervise agriculture due to livelihood supply. Religion takes place from one place to another among the Bhils. Most of them worship native deities, such as Khandoba, Kanhoba, Bahiroba, and Sheetalamata. Some worship was done by Tiger God. He has no temple of his own. The Bhils are highly superstitious tribal people. They have Bhagat or Guru perform UN agency spiritual rites. He has created cultural history and lends abundant importance to bounce and music. Ghoomar is the most known dance among the Bhils. Then there is the religious dance drama performed by men in the month of Gair Shravan (July and August). The Bhils are talented in sculptural work. They make beautiful horses, elephants, tigers, gods from the soil. Baneshwar is honest that the main festival is celebrated among the Bhils. This is for the entire amount of honest Shivaratri (January or February) and furthermore Baneeshwar Mahadev is called Lord Shiva. On this occasion, the Bhils all camp together on the banks of the Som and Mahi River. They dance on the stove and sing ancient songs. At night, all Hindu deities enjoy Raslila in the Narayana temple. Cultural shows, magic shows, and animal shows are the major attractions of the Kalabaz feast fair. This fair is a combination of two fairs, which are held in reverence to Lord Shiva and another that begins after the setting up of Vishnu Temple by Jankunwari. The sacred and Dussehra class measures the major festivals celebrated among the Bhils in Asian countries. They dance with fire and sing traditional songs. At night, they all enjoy Raslila at the Lakshmi Narayan Temple. This fair is a combination of two fairs, which are held in reverence to Lord Shiva and another that begins after the setting up of Vishnu Temple by Jankunwari. On this occasion, the Bhils all camp together on the banks of the river Som and Mahi. Their men's and women's costumes are unique. Bhil women wear an upper garment known as Odhana, Ghaghra, and Kapda. In the olden times, women of the Bhil community used to facilitate the small size of the skirt (knee length) to be repeatedly shaken. The variety of colors also make tribal costumes more attractive. At a similar end, an ornament named Pejania is worn on the arms, legs, and hands that protect animals and thorns. Women cover their heads and torso with Lugada or Odhna. They use beautiful jewelry that matches their ethnicity and dress. Some of the popular jewelry worn by women in the Bhil community include Bichhiya, Dhimna, Binti, Ognia, Pejania, Hansali, Kasla, Hara, Beedi, Tagli, and Kamakada. They use white metal and silver brass in jewelry. The men of the Bhil community wear turban, angarkha, angi, and potario (a lower cloth).

9.4.2 SAHARIYA

The Sahariya tribe is a primitive tribe group only in Rajasthan. They are concentrated in the Shahabad and Kishanganj panchayat committees of Baran district. Sahariya is backward and one of the first settlers of Rajasthan. Tod (1881) mentions them along with Minas, the Bhils, and Gurjars because of the primitive inhabitants of the area. The word Sahariya is derived from the Persian world, "sehi" meaning jungle. Muslim rulers considered Saharia as the inhabitants of the forest. The customs and methods of the Saharas bear a great resemblance to the caste Hindus, with whom they live in their present abode and who consider them untouchables. The costumes of Sahariya men were of a short length—dhoti, saluk (shirt), and turban. Sahariya women wore ghagra, lame, and choli (choli). The young boys wore panch (small piece of dhoti) and saluk and the girl wore clothes of Ghaghria, Palka, and Faria. Very young children went without clothes and were seen naked. Saharis usually live in a separate settlement in the village called Saharan. They follow Hindu nonsecular practices and speak utterances influenced by Hadoti. Cohorts maintained ecological balance with their environment for ages, despite the low level of technology. Mostly they were out of marginal forest and engaged as agricultural laborers. The main business is collecting and selling forest wood, gum, tendu leaf, honey, fruits, and vegetables. Although Saharians are normally primitive, they are not all like that. Some of them are also inhabited farmers. Others are landless laborers and forest producers. Sitabari fair is famous. The holy place Sitabari is about 45 km from Baran town near Kelwara town. A large social group is sincere on the first half of the month. There are many pools for Sita, Lakshmana, Surya, Luv, and Kush at this place, and taking a holy dip in them is considered auspicious. Millions of people return here for this honesty taken within the Sahariya tribe as Kumbh. The Swayambara (marriage ceremony) of Sahariya tribes are unionized during this honest, where people from the state of Rajasthan come and are associated with Madhya Pradesh, Sahariya boy drops a handkerchief to propose Sahariya girl at the fair. If the Saharia girl accepts a handkerchief, it is said that they agree to the marriage. The bride and groom take seven rounds of the Barnava tree (sitting here), and after the blessings of their parents, they are accepted as a married couple. This honest is additionally referred to as animal honest where there is a sense of selling, buying and selling of cows, buffalo, and so on. This social group provides a perfect opportunity to observe life trends.

9.5 SOME INSIGHT

Rajasthan is very famous for its cultural activities and religious places where people follow unique cultures and traditions. Baneshwar fair is a popular tribal fair, which is a religious fair with simple and traditional rituals. The Bhil's tribals come from neighboring states of Madhya Pradesh and Gujarat and offer prayers to Lord Shiva. This fair showcases all the fun and charm of fair life and sells local goods. The fair showcases songs, folk dances, magic shows, animal shows, and acrobatics.

There is a huge procession of priests called the Matadhish, reaching the fair site. Although bathing the Mathadhish, the water of the river is purified purportedly. Therefore, individuals bathe him within the stream. Sitabari Honest, the command at Sitabari is to commemorate the sanctity of the place and its nonsecular and historical significance as an annual upright. The honest Sitabari at Sitabari reflects the nonsecular importance of the city of Sitabari within Hinduism. Absolute honest celebrations create honest imagery with a range of events, fun, and engaging activities during individuals. The Swayambar ceremony is held within the upright, which people do a lot with their relationships. Honest Bundi, persons from areas like Nagaur, Aklera, Jhalawar, the Bhilwara, Kota, and so on, are the best reasons for the attraction of Sitabari city. Both fairs have a large number of shops in the fair for purchase and sale of essential items and fancy articles. Beautiful handmade and readymade products, shops, and stalls are also available at the fair at very low prices. Cultural programs are organized by the children of the tribe. Groups of villagers are invited to participate in the program. All these activities are a boon for the economic development of these villagers. Keeping this in mind, the present chapter is organized. Two fairs of two different tribes of the same state served for further research. The Bhil community is socially and economically better than the Saharis. After visiting such villages, it was felt that poverty and illiteracy were the cause of lack of regular source of income. They are communities that feel isolated and isolated from the rest of society. They believe in their traditions and customs and never think beyond it. But some young tribal groups who have already performed as folk artists in many venues took interest to make their community more educated and civilized. Previous studies have already shown (Ryan and Aicken, 2005) that there are two aspects of tribal tourism development, that is, capacity building and commodification. Capacity building is one of the major positive aspects and the other is what reflects their lifestyle for tourism development. The present study further offers tribal tourism fairs in place of tribal tourism and tribal fairs. This can eliminate the ill effects of the amendment and increase the priorities of the source of income. Researchers interviewed villagers, spoke about the development of tribal fairs. They responded in a very positive way but they never took their religious customs or fairs with economic benefit. Most villagers consider their fairs to be social or religious customs. Although many of them were financially involved with these fairs, they took the fairs in their own way for economic gain.

9.6 CONCLUSION

The tribal fair has great potential to develop tourism and can have a positive impact on local livelihoods. A strategic intervention is needed to ensure the commercial viability of folk cultures and traditions. It is one of the major ways to empower tribes and their artists from a socioeconomic point of view. All possible efforts should be made to maintain the innate beauty, inherent quality, and core ideology of tribal/folk art and cultural forms. This is necessary to ensure their exclusive uniqueness. The general public of the wider society should be informed about the values, styles, forms of various folk performing arts. Effective campaigning and publicity of these elements are to crystallize favorable public attitudes and ideas toward folk art and culture; the pattern of presentation of these tribal/folk art and cultural forms should be revised and strategically altered to make them more vibrant to meet the recreational demand of the present generation. This chapter concludes that a tribal culture is a tool for their economic and social development, but the strongest tool for their overall development in their religious and social fairs. Through these fairs, their handicrafts, customs, dishes, dishes, and other articles can be an exhibition and sold to a large number of people. Such events are forced community participation and core involvement. That way they will not feel commodified or awkward. They will be proud of their religious and social customs, customs, and traditions.

KEYWORDS

- **tribal tourism**
- **tribal fair**
- **Rajasthan**
- **cultural**

REFERENCES

Butler, R.; Hinch, T. *Tourism and Indigenous Peoples*; Routledge: Abingdon, UK, 2007.

Carr, A.; Ruhanen, L.; Whitford, M. Indigenous Peoples and Tourism: Challenges and Opportunities for Sustainable Tourism. *J Sustain. Tour.* **2016,** *24* (8–9), 1067–1079.

Chang, J. Segmenting Tourists to Aboriginal Cultural Festivals: An Example in the Rukai Tribal Area, Taiwan. *Tour. Manage.* **2006,** *27* (6), 1224–1234.

Chang, H.-M.; Taiwan Associates; Pro, L.-C.; Liao, C. A Study of Indigenous Tribe Tourism Planning and Developing—Case by Huanshan in Taiwan. *J. Int. Manage. Stud.* **2009,** 9 (1), 146–155.

Douglas, N.; Douglas, N.; Derrett, R. *Special Interest Tourism*; John Wiley and Sons: Milton, QLD, 2001. ISBN: 0471421715.

Getz, D.; Jamieson, W. *Rural Tourism in Canada: Issues, Opportunities and Entrepreneurship in Original Tourism in Canada*; Cengage Learning EMEA: Alberta, Canada, 1997; pp 93–107.

Heather, Z. D. *Indigenous Ecotourism: Sustainable Development and Management*; CABI: Wallingford, UK, 2006.

Liao, C.-C.; Lin Y.-X. A Study of Indigenous Tribes Tourism Developing Case by Lilang, Tbulan, and Hrung in Taiwan, World Academy of Science, Engineering and Technology. *Int. J. Econ. Manage. Eng.* **2013,** *7* (11), 2945–2951.

Notzke, C. Indigenous Tourism Development in the Arctic. Ann. Tour. Res. **1999,** *26* (1), 55–76.

Ryan, C.; Aicken, M. Tourist-Host Nexus—Research Considerations. In *Indigenous Tourism: The Commodification and Management of Culture. Advances in Tourism Research*; Elsevier, Amsterdam, 2005; pp 1–11.

Ryan, C.; Houghton, J. Tourists and Aboriginal People. *Ann. Tour. Res.* **2002,** *29* (3), 631–647.

Smith, V. L. The Four *H*s of Tribal Tourism: Acoma—A Pueblo Case Study. *Progress Tour. Hosp. Res.* **1996,** 2 (3–4), 295–306.

Vidyarthi, L. P.; Rai, B. K. *The Tribal Culture of India*; Concept Publishing Company: Delhi, 1977.

Ward, T. *Tribal Tourism*; 2010. https://www.aol.com/2010/12/08/tribal-tourism-ethical-or-exploitative/.

Weiler, B.; Hall, C. M. In *Special Interest Tourism*; Weiler, B.; Hall, C. M., Eds.; Belhaven Press/Halsted Press: New York, 1992.

Whitford, M.; Ruhanen, L. Indigenous Tourism Research, Past and Present: Where to from Here? *J. Sustain. Tour.* **2016,** *24* (8–9), 1080–1099.

INDEX

D

G

H

I

K

R